N
FRA   G

Matches t

Queenswood

NAME	FORM	DATE
Andrea Ball	9x	7.9.09
Anais Lang	9y	9/9/10
Olivia Wilson	9Y	8/9/17

Kevin Evans, Keith Gordon, Trevor Senior, Brian Speed

Property of Queenswood School
Mathematics Department

Contents

CHAPTER 1 Algebra **1 & 2** 1

CHAPTER 2 Number **1** 6

CHAPTER 3 Algebra **3** 16

CHAPTER 4 Geometry and Measures **1** 21

CHAPTER 5 Statistics **1** 28

CHAPTER 6 Geometry and Measures **2** 36

CHAPTER 7 Number **2** 42

CHAPTER 8 Algebra **4** 49

CHAPTER 9 Statistics **2** 53

CHAPTER 10	Geometry and Measures **3**	58
CHAPTER 11	Algebra **5**	63
CHAPTER 12	Solving Problems and revision	67
CHAPTER 13	Statistics **3**	83
CHAPTER 14	Geometry and Measures **4**	85
CHAPTER 15	Statistics **4**	88
CHAPTER 16	GCSE Preparation	90

Introduction

Welcome to *New Maths Frameworking*!

New Maths Frameworking Year 9 Practice Book 3 has hundreds of levelled questions to help you practise Maths at Levels 6-8. The questions correspond to topics covered in Year 9 Pupil Book 3 giving you lots of extra practice.

These are the key features:

- **Colour-coded National Curriculum levels** for all the questions show you what level you are working at so you can easily track your progress and see how to get to the next level.

- **Functional Maths** is all about how people use Maths in everyday life. Look out for the Functional Maths icon **FM** which shows you when you are practising your Functional Maths skills.

CHAPTER 1 Algebra 1 & 2

Practice

1A Sequences

1 Find the next three terms in these sequences.
  a 4, 11, 18, 25 ...
  b 100, 92, 84, 76 ...
  c −25, −19, −13, −7 ...
  d 2.43, 2.54, 2.65, 2.76 ...
  e 3, 5, 8, 12, 17 ...
  f 1, 3, 7, 13, 21 ...

2 Write down the first four terms of the sequences with these nth terms.
  a $6n + 2$
  b $3n - 10$
  c $-2n + 3$
  d $\frac{n+1}{2}$
  e $0.7n$
  f $2.3n - 0.4$

3 Find the nth term for each of these sequences.
  a 5, 10, 15, 20 ...
  b 2, 3, 4, 5 ...
  c 7, 10, 13, 16 ...
  d 8, 18, 28, 38 ...
  e −3, −6, −9, −12 ...
  f −10, −8, −6, −4 ...
  g 0.3, 0.6, 0.9, 1.2 ...
  h 6.6, 7.4, 8.2, 9 ...
  i $\frac{3}{2}, \frac{6}{3}, \frac{9}{4}, \frac{12}{5}$ ...

4 Find the nth term for each of these sequences.
  a 25, 24, 23, 22 ...
  b 12, 10, 8, 6 ...
  c 100, 95, 90, 85 ...
  d −4, −8, −12, −16 ...
  e −5, −8, −11, −14 ...
  f 7.2, 7, 6.8, 6.6 ...

5 The tables show Martha and Johan's bank balances during the first five weeks of 2003.

Johan's bank account

Week	1	2	3	4	5
Balance, £	970	1035	1100	1165	1230

Martha's bank account

Week	1	2	3	4	5
Balance, £	340	290	240	190	140

  a Write a formula to show the balance of Johan's bank account after n weeks of 2008.
  b Write a formula to show the balance of Martha's bank account after n weeks of 2008.
  c What will be the balance of **i** Johan's **ii** Martha's bank account after nine weeks of 2008?

Martha and Johan combine their bank balances.
  d Make a new table to show their combined bank balance during the first five weeks of 2008.
  e Write a formula to show their combined balance after n weeks of 2008.
  f Martha and Johan are saving for a holiday that costs £2200. When will they have saved enough?

Practice
1B Quadratic sequences

 This diagram shows a growing pattern of equilateral mosaic tiles.

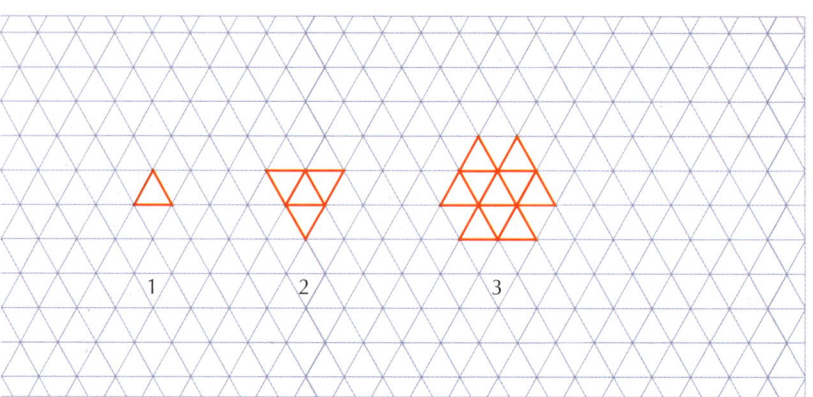

Each new pattern is made by placing a tile alongside every edge of the previous pattern.

a Use triangular grid paper to draw the next pattern.
b Copy and complete this table.

Pattern number, n	1	2	3	4
Number of tiles in pattern	1	4		

c Describe a rule for finding the number of tiles in a pattern.
d i Use your rule to calculate the number of tiles in the fifth pattern.
  ii Verify your rule works by drawing the fifth pattern.

 This diagram shows how different numbers of flowerpots can be arranged in a row.

1 pot
1 arrangement

2 pots
2 possible arrangements

3 pots
6 possible arrangements

a Find the number of ways four flowerpots can be arranged in a row.
b Copy and complete this table.

Number of flower pots, n	1	2	3	4
Number of ways flower pots can be arranged in a row				

c Describe a rule for finding the number of ways n flowerpots can be arranged in a row.
d i Use your rule to calculate the number of ways five flower pots can be arranged.
  ii Verify that your rule works by drawing a diagram.
    Hint: You don't need to draw all of the arrangements.

3 Write down the first four terms of the sequences with these nth terms.
  a $n^2 + 4$ **b** $2n^2 + 3n$ **c** $5n^2 + 2n - 3$

Practice

1C Functions

1 Find the inverse of each function.
  a $x \to 6x$ **b** $x \to x + 7$ **c** $x \to \frac{x}{3}$ **d** $y = x - 1.5$

2 Find the inverse of each function.
  a $x \to 4x - 3$ **b** $x \to 10x + 2$ **c** $y = 2x - 4$ **d** $x \to \frac{x}{3} + 5$

3 Find the inverse of each function.
  a $x \to 4(x + 2)$ **b** $y = 3(x - 5)$ **c** $x \to \frac{(x + 4)}{3}$

4 i Find the function for each mapping diagram.
  ii Find the inverse of the function.

  a $x \to ?$
  1 → 4
  2 → 8
  3 → 12
  4 → 16

  b $x \to ?$
  3 → 6
  4 → 7
  5 → 8
  6 → 9

  c $x \to ?$
  1 → 3
  2 → 5
  3 → 7
  4 → 9

5 Which of these functions are **i** identities and **ii** self-inverse functions?
  a $x \to 2x - 2$ **b** $x \to -10 - x$ **c** $x \to \frac{3x}{3}$ **d** $x \to \frac{2}{x}$

6 Each mapping diagram is of the inverse of a function.
  i Find the inverse function.
  ii Find the original function.

  a $? \leftarrow x$
  1 ← 6
  2 ← 7
  3 ← 8
  4 ← 9

  b $? \leftarrow x$
  1 ← 5
  2 ← 8
  3 ← 11
  4 ← 14

Practice
1D Graphs

 1 Ben and Gurjit left their homes at the same time and travelled to each other's home. This graph shows their journeys.

Give reasons for your answers to these questions.

a One person walked and ran, the other cycled. Which person cycled?

b Gurjit and Ben met each other.
  i When was this?
  ii How far were they from Gurjit's home?
  iii How long was their meeting?

c There is a hill between their two homes. Roughly how far from Ben's home is the top of the hill?

d When did Ben travel slowest?

e When did Gurjit travel fastest?

f Ben and Gurjit both started their return journeys at 9.30am. Ben waited one minute at the top of the hill for Gurjit to arrive. They chatted for 2 minutes. Sketch their return journeys.

 2 Each of these candles is 25 cm tall and burns to the ground in 20 hours.

a 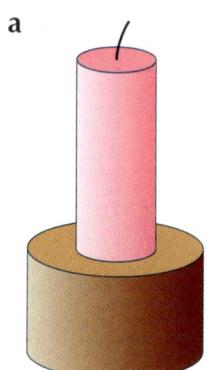   b    c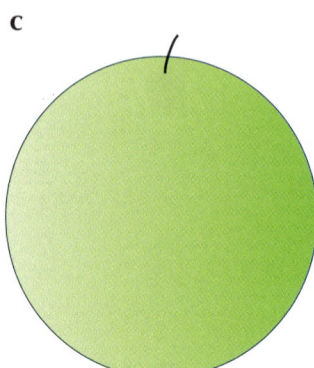

For each candle, sketch a graph showing its height over time.

3 Match each graph with the correct description.

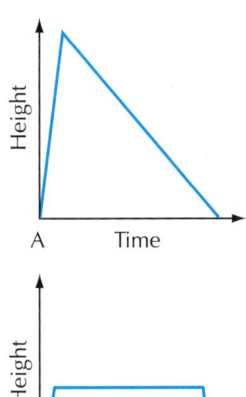
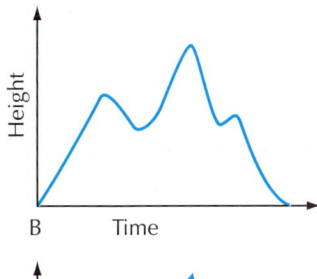
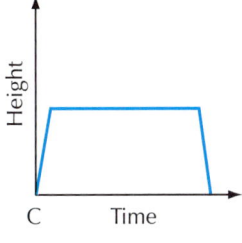
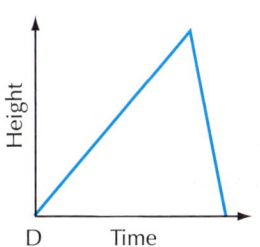

a A hovercraft journey.
b A rocket launch and parachute landing.
c A flight of a radio-controlled model aeroplane.
d A model hot air balloon that takes off and catches fire.

4 Sketch a graph to show these.

a The height of a space shuttle that takes off, circles the Earth and lands.
b The weight of a woman 3 months before she is pregnant, during her pregnancy and 3 months after she gives birth.

Practice
1E Limits of sequences

1 a Calculate the first eight terms of this sequence:
   First term: 1
   Term-to-term rule: divide by 4 then add 6
   b State the limit of the sequence.

2 a Calculate the first eight terms of this sequence:
   First term: 1
   Term-to-term rule: divide by 5 then add 10
   b State the limit of the sequence.

3 a Copy and continue these calculations to give the first eight terms of a sequence.

$$1 = 1$$
$$1 + \tfrac{1}{2} = 1.5$$
$$1 + \tfrac{1}{2} + \tfrac{1}{4} =$$
$$1 + \tfrac{1}{2} + \tfrac{1}{4} + \tfrac{1}{8} =$$
$$1 + \tfrac{1}{2} + \tfrac{1}{4} + \tfrac{1}{8} + \tfrac{1}{16} =$$

b State the limit of the sequence, if there is one.

4 a Calculate, as decimals, the 10th, 100th and 1000th terms of the sequence $\frac{1}{2}, \frac{2}{3}, \frac{3}{4}, \frac{4}{5}$ ...
b State the limit of this sequence.

5 a Calculate, as decimals, the first ten terms of the sequence whose *n*th term is $10 - \frac{1}{n}$.
b Calculate, as decimals, the 100th and 1000th terms.
c State the limit of the sequence.

6 Find a sequence whose limit is 4. Write down its *n*th term.

Hint: Look at Question 3.

7 Use a spreadsheet to investigate the limit of the sequence produced by these calculations.

$1 \quad =$

$1 + \frac{1}{2} \quad =$

$1 + \frac{1}{2} + \frac{1}{3} + \frac{1}{4} \quad =$

$1 + \frac{1}{2} + \frac{1}{3} + \frac{1}{4} + \frac{1}{5} \quad =$

etc.

CHAPTER 2 Number 1

Practice

2A The four rules governing fractions

Convert these fractions to equivalent fractions with the same denominator. Work out the answer. Cancel your answer and write as a mixed number, if necessary.

1 a $2\frac{1}{3} + 1\frac{2}{5}$ b $3\frac{1}{4} + 4\frac{2}{3}$ c $1\frac{5}{6} + 2\frac{2}{3}$
d $3\frac{7}{9} + 6\frac{1}{6}$ e $7\frac{1}{6} + 5\frac{3}{4}$ f $5\frac{3}{8} + 4\frac{1}{10}$

2 a $3\frac{3}{4} - 1\frac{2}{3}$ b $5\frac{5}{6} - 2\frac{4}{5}$ c $4\frac{1}{8} - 1\frac{1}{3}$
d $6\frac{5}{6} - 3\frac{1}{4}$ e $5\frac{5}{8} - 1\frac{9}{10}$ f $9\frac{1}{6} - 4\frac{7}{9}$

3 Calculate each of the following.

a $\frac{5}{6} - \frac{7}{15}$ b $\frac{5}{16} + \frac{13}{20}$ c $\frac{13}{18} - \frac{7}{12}$
d $\frac{9}{10} + \frac{7}{15}$ e $\frac{13}{14} - \frac{3}{8}$ f $\frac{11}{12} - \frac{13}{24}$

4 Marion buys a tub containing $7\frac{7}{8}$ oz of peanuts and a packet containing $3\frac{5}{6}$ oz of peanuts.

   a What is the total weight of the peanuts she bought?
   b How much more does the tub contain than the packet?

Cancel before multiplying, if possible. Cancel your answer and write as a mixed number, if necessary.

5 **a** $\frac{4}{9} \times \frac{2}{5}$ **b** $\frac{5}{8} \times \frac{2}{3}$ **c** $\frac{8}{15} \times \frac{5}{7}$
   d $\frac{10}{21} \times \frac{7}{15}$ **e** $\frac{9}{16} \times \frac{12}{27}$ **f** $\frac{8}{9} \times \frac{6}{15} \times \frac{5}{12}$

6 Write as improper (top-heavy) fractions first.
   a $1\frac{2}{3} \times 1\frac{3}{4}$ **b** $2\frac{4}{5} \times 1\frac{2}{7}$ **c** $2\frac{5}{8} \times 3\frac{5}{9}$
   d $4\frac{4}{5} \times 3\frac{1}{3}$ **e** $2\frac{7}{10} \times 2\frac{7}{9}$

7 **a** $\frac{5}{8} \div \frac{2}{3}$ **b** $\frac{3}{5} \div \frac{7}{10}$ **c** $\frac{15}{16} \div \frac{9}{10}$
   d $\frac{8}{9} \div \frac{10}{21}$ **e** $\frac{16}{25} \div \frac{14}{15}$

8 Write as improper (top-heavy) fractions first.
   a $2\frac{2}{3} \div 1\frac{2}{5}$ **b** $1\frac{1}{6} \div 2\frac{3}{4}$ **c** $3\frac{5}{9} \div 3\frac{2}{3}$
   d $4\frac{1}{8} \div \frac{33}{40}$ **e** $\frac{8}{15} \div 5\frac{3}{5}$

9 A bottle contains $7\frac{7}{8}$ ounces of perfume. A perfume spray contains $\frac{7}{10}$ ounces of perfume.

   a How many sprays can be filled from the bottle?
   b How much perfume is left over?

Practice
2B Percentages and compound interest

1 Write down the multiplier equivalent to each of these.

   a 15% increase   **b** 2% decrease   **c** 95% increase
   d 60% decrease   **e** $12\frac{1}{2}$% increase   **f** $4\frac{1}{4}$% decrease
   g 33.3% increase   **h** 8.7% decrease   **i** $17\frac{1}{2}$% increase

2 How much would you have in the bank if you invested:

   a £500 at 4% interest for 3 years?
   b £8000 at $3\frac{1}{2}$% interest for 5 years?
   c £250 at 5.3% for 10 years?

3 In 1960, 8000 cars were abandoned in a city. Each year, 8% more cars were abandoned.

  a How many cars were abandoned in 1966?
  b Which was the first year when more than 20 000 cars were abandoned?
  c How many cars were abandoned in 2000?

4 A snowman has a volume of 600 litres. Each day, it loses 13% of its volume.

  a What will its volume be after 4 days?
  b How many days will it take for there to be less than 100 litres of snow left?
  c What will be its volume after 20 days?

Practice 2C Reverse percentages and percentage change

1 a From 1980 to 1998, the Royal Air Force increased its membership by 12% to 46 900. How many people were in the Royal Air Force in 1980?
  b During the same period, the Royal Navy decreased its size by 17% to 28 500. How many people were in the Royal Navy in 1980?

2 What were the prices before VAT was added to these items?

a  b 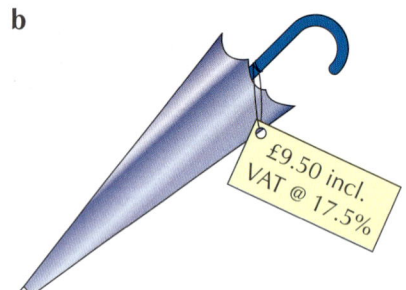 c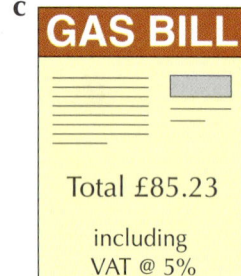

3 In one year, a eucalyptus tree grew from 2.7 m tall to 4.3 m tall. What was the percentage increase in height?

4 Frances filled her petrol tank with 60 litres of petrol. After a journey, there were 18 litres left. What was the percentage decrease in fuel?

5 This table shows how the populations of some towns and villages changed between 1980 and 2000. Copy and complete the table.

Place	Percentage change	Population in 1980	Population in 2000
Smallville	30% decrease		1300
Lansbury	8% increase	46 000	
Gravelton	2% decrease		19 800
Smithchurch	48% increase		144 000
Deanton		28 000	33 320
Tanwich		680	646

Practice
2D Direct and inverse proportion

1 Martha takes 30 minutes to read 20 pages.
How long will it take her to read 30 pages?

2 6 boxes contain 54 light bulbs.
How many light bulbs are contained in 9 boxes?

3 Lai Ping has 10 rows of 18 football stickers on his bedroom wall.
He rearranges them into 12 rows. How many stickers are in each row?

4 9 people take 24 hours to sew a patchwork quilt.
   a How long would 6 people take?
   b How many people would be needed to finish the quilt in 4 hours?

5 The diagram shows part of a square tiled floor.

   a If the floor has 54 black tiles, how many white tiles does it have?
   b If the floor has 400 tiles altogether, how many white tiles does it have?

6 A packet of grass seed can cover a rectangular lawn measuring 6 metres by 9 metres. Maria uses a whole packet on her lawn that is 4 metres wide. How long is her lawn?

7 A new battery can run 3 model trains for 8 hours.
How long can it run 6 trains for?

8 5 identical candles lit at the same time take 60 minutes to burn down.
How long will 9 candles take?

9 a 18 eggs cost a shopkeeper £2.88. How much do 10 cost?
   b 60 egg boxes cost £3.90. How much do 150 cost?

10 A value pack of sweets contains 6 Tweeny Bars, 8 Swirlies and 12 Fruitgums.

Alma bought some value packs for a party. She bought 32 Swirlies altogether.
   a How many Fruitgums did she buy?
   b How many Tweeny bars did she buy?

Jan bought value packs containing 182 sweets altogether.
   c How many Swirlies did he buy?

11 During a maths lesson for 15 pupils, the fire alarm sounded for 20 seconds. How long will the fire alarm sound when there are 20 pupils in the class?

12 An aeroplane journey took 2 hours at an average speed of 300 mph.

  a How long would it take to travel the same journey at:
   i 200 mph?   **ii** 400 mph?   **iii** 900 mph?
   Give your answers in hours and minutes.
  b How fast would the aeroplane need to travel to complete the journey in:
   i hour?   **ii** 5 hours?   **iii** 45 minutes?
   Hint: Convert minutes to hours first.

13 Jason used 6 litres of paint to cover 15 m of fence.
How much paint is needed to cover 40 m of fence?

14 6 parachutists jump out of an aeroplane and take 12 minutes to land.
How long will 8 parachutists take to land?

15 It takes 6 bricklayers 3 hours to build a wall 100 cm high.

  a How long would 4 bricklayers take to build the same wall?
  b How long would it take 6 bricklayers to build a wall 200 cm high?
  c How many bricklayers would it take to build a 180 cm wall in 2 hours?

16 A tortoise and a hare had a race. They travelled at a constant speed. When the tortoise had travelled 24 metres, the hare had travelled 108 metres.

  a How far had the tortoise travelled when the hare had travelled 126 metres?
  b How far had the hare travelled when the tortoise had travelled 100 metres?
  c If the hare took 4 minutes to travel 108 metres, how long did the tortoise take?

Practice 2E Ratio in area and volume

1 The diagram shows pairs of similar shapes.

  a

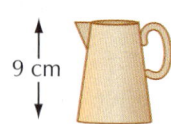

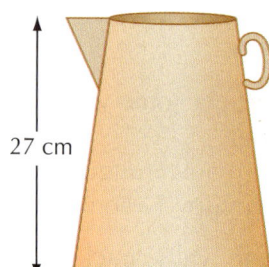

9 cm

27 cm

b

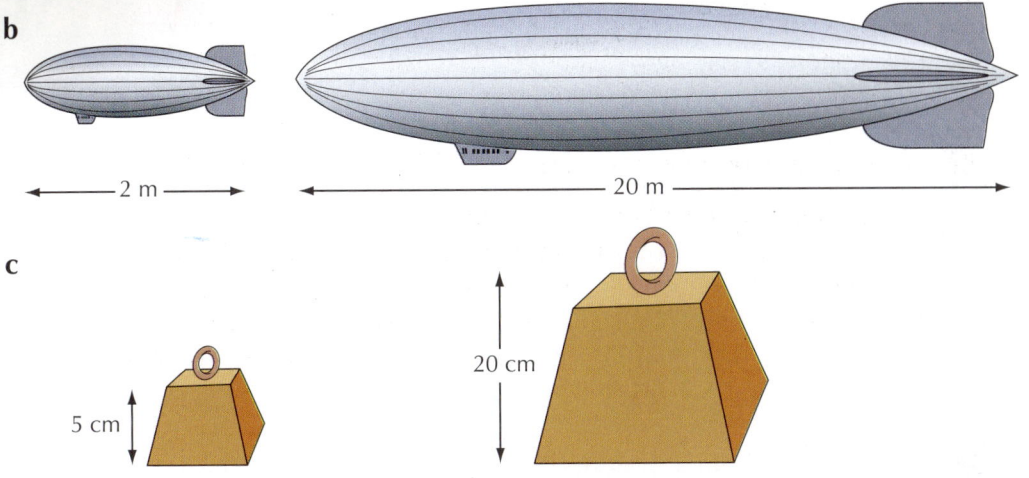

c

For each pair, calculate the ratio of the:

i lengths.　　　　ii areas.　　　　iii volumes.

Give your answers in the form $1 : n$.

2 The diagram shows pairs of similar shapes. Find the missing areas.

a

b

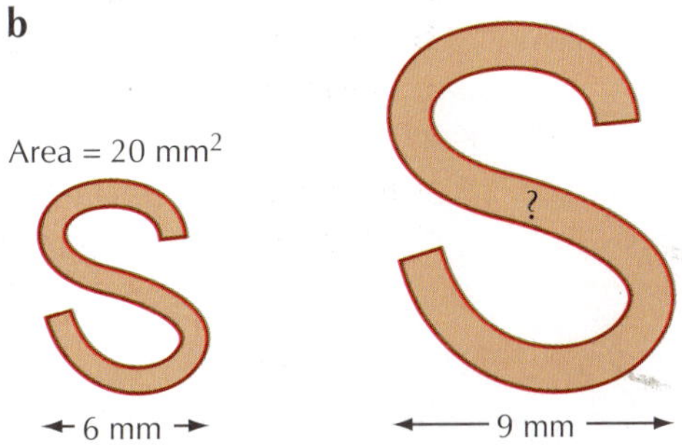

c

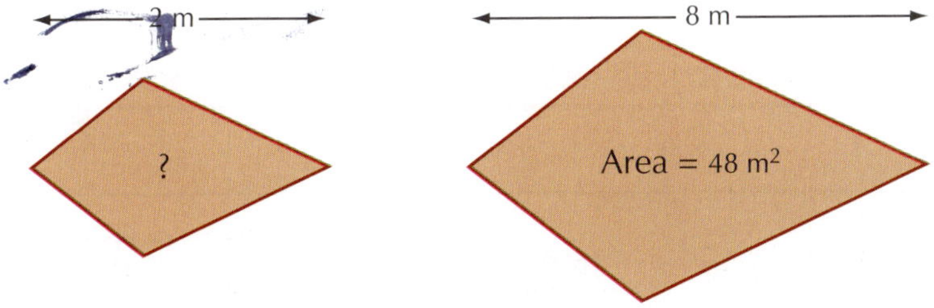

3 The diagram shows pairs of similar shapes. Find the missing volumes.

a

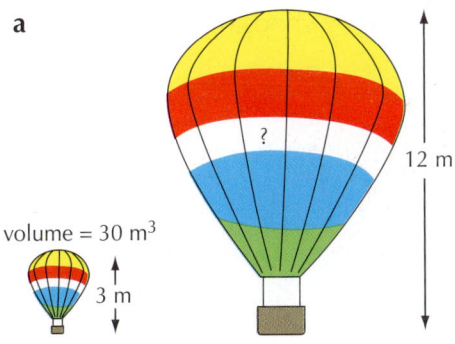

b

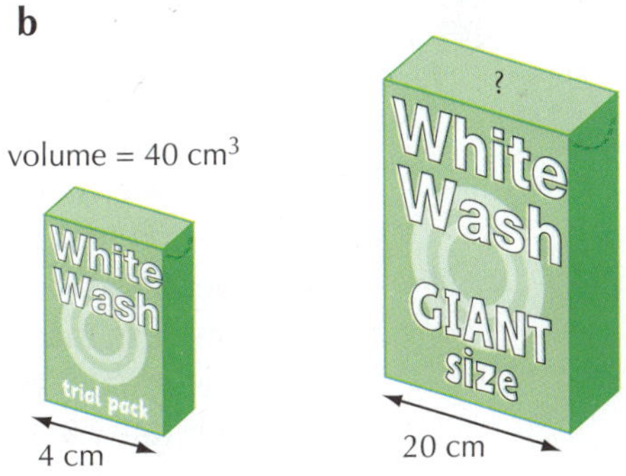

c

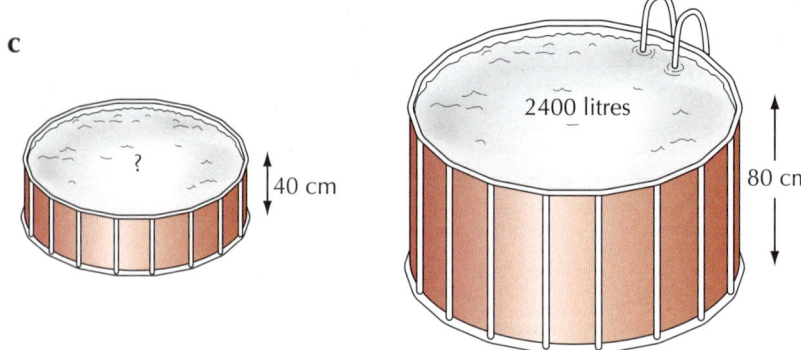

4. These tiles are similar to each other. Find the missing areas, volumes and thickness.

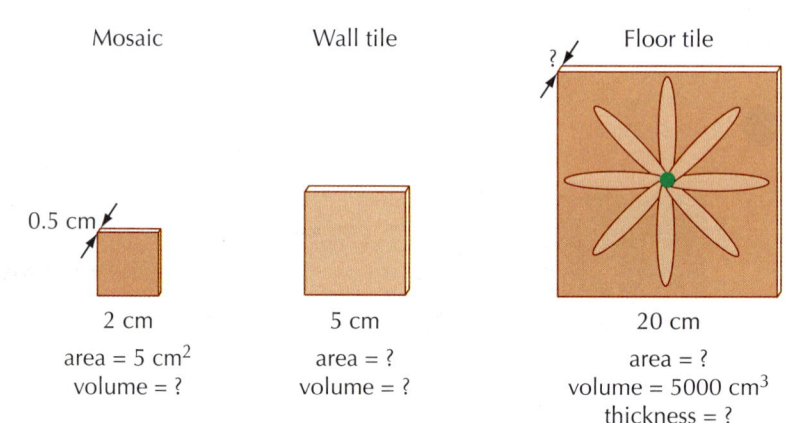

Practice

2F Numbers between 0 and 1

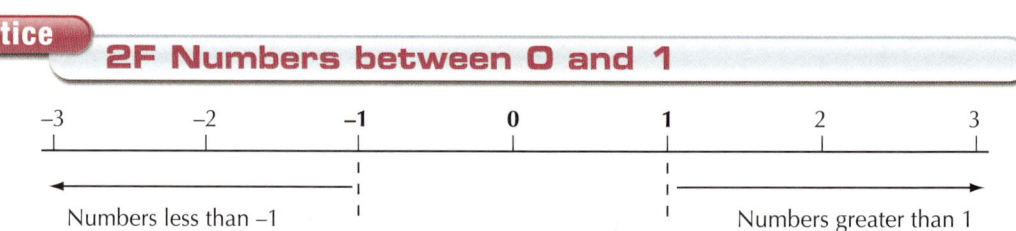

Use your calculator to help you to answer these questions.

1. Describe the result of each calculation, **a** to **h**, by choosing one of these answers:

   A number between −1 and 0
   A number between 0 and 1
   A number greater than 1
   A number less than −1
   A number less than 0
   A number greater than 0
   A number between −1 and 1

0
1
−1
Impossible
None of the above

a Squaring a number greater than 1
b Finding the square root of a number between 0 and 1
c Squaring −1
d Squaring a number less than −1
e Finding the square root of a number between −1 and 0
f Dividing −1 by a number less than −1
g Dividing 1 by a number between −1 and 0
h Dividing 0 by 0

2 a Describe the effect of multiplying any number by −1.
  b Describe the effect of dividing any number by −1.

3 Describe the effect of dividing 1 by a number between −1 and 1.

4 a
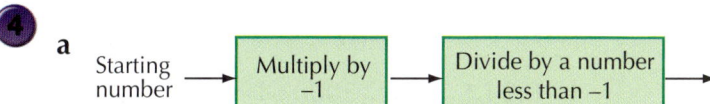

Work your way along the flow chart.
Describe the result using one of the answers shown in Question 1.

  i   Start with a number between 0 and 1.
  ii Start with −1.
  iii Start with a number between −1 and 1.

b Repeat part **a** with this flowchart.
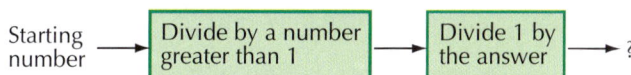

Practice

2G Reciprocal of a number

Do not use your calculator for the first five questions. Show your working.

1 Calculate the reciprocal of each number. Give your answer as a decimal or whole number.

a 2   b 10   c 0.1   d 4   e −1
f 0.2   g −5   h 20   i −1000 j 25

2 Calculate the reciprocal of each fraction. Give your answer as a fraction or whole number.

a $\frac{1}{6}$ b $\frac{3}{4}$ c $\frac{1}{20}$ d $\frac{10}{9}$ e $1\frac{2}{3}$
f $\frac{3}{100}$ g $-\frac{1}{8}$ h $4\frac{4}{5}$

3. Complete these. Work in whole numbers and fractions.
   a square of 3 =
      reciprocal of square of 3 =
   b reciprocal of 3 =
      square of reciprocal of 3 =
   c the reciprocal of the square of a number =

Use a calculator to help you answer these questions.

4. Calculate the reciprocal of these using the reciprocal key on your calculator. Round your answers appropriately.
   a 40   b 0.7   c 125   d 0.003
   e 3.2   f $3\frac{1}{8}$ g −4.5   h 10^5

5. Convert the ratios to the form 1 : n, using the following example of working.

   $$50 : 1$$
   Divide both sides by 50. $50 \div 50 : 1 \div 50$
   $$1 : 0.02$$

   a 5 : 1   b 0.8 : 1   c 2000 : 1   d $2\frac{1}{2}$: 1
   e 12 : 1   f 0.025 : 1   g $\frac{1}{7}$: 1   h $5\frac{2}{3}$: 1

Practice — 2H Rounding and estimation

1. Round these numbers **i** to 1 decimal place and **ii** to 2 decimal places.
   a 2.874   b 0.05537   c 20.957
   d 0.5059   e 9.98407

2. Round these numbers **i** to 1 significant figure and **ii** to 2 significant figures.
   a 327   b 8.452   c 8175   d 0.06594
   e 79 550   f 0.007 29   g 20.981

3. Round these quantities to an appropriate degree of accuracy.
   a Jeremy is 1.8256 m tall.
   b A computer disk holds 688 332 800 bytes of information.
   c A petrol tank has a capacity of 72.892 litres.
   d An English dictionary contains 59 238 722 words.
   e Kailash held his breath for 82.71 seconds.

4 Calculate these.

   a 400 × 60   b 2000 × 900   c 0.4 × 0.8
   d 0.02 × 0.7   e 0.04 × 0.09   f 300 × 0.2
   g 0.08 × 4000 h 60 × 0.9

5 Calculate these.

   a 800 ÷ 20   b 12 000 ÷ 300   c 0.8 ÷ 0.4
   d 0.12 ÷ 0.03   e 6 ÷ 0.3   f 40 ÷ 0.05
   g 3000 ÷ 0.6

6 Estimate answers to these by rounding the numbers to 1 significant figure first. Show your working.

   a 278 × 59   b 9483 ÷ 34   c 0.358 × 166
   d 43 ÷ 0.537   e 0.0366 × 0.0899   f 777 ÷ 0.241
   g 3278^2 h 2.67 + 5.18 × 9.01 i 53 × (0.672 − 0.306)

7 Estimate answers to these.

   a $\dfrac{5.018 \times 3.86}{11.41 - 6.459}$ b $\dfrac{37.71 + 18.099}{0.829 - 0.333}$

   c $\dfrac{57}{0.19^2}$ d $\dfrac{87^2}{0.0343 \times 2480}$

8 This table shows the prices of some metals.

Metal	Price
Lead	£0.2412 per kg
Silver	£0.085 per g
Gold	£2575 per kg
Platinum	£11.63 per g

   a Estimate the cost of buying these. Show your working.
      i   2474 kg of lead   ii   0.2856 g of silver
      iii 0.036 kg of gold   iv   19.08 g of platinum
   b Estimate answers to these. Work in pounds. Show your working.
      i   How much gold could be bought for £64 250?
      ii   How much lead could be bought for £124.59?
      iii How much platinum could be bought for £289 320?
      iv   How much silver could be bought for £0.94?

CHAPTER 3 Algebra 3

Practice

3A Equations and formulae

For questions 1–2, solve the equations. Show your working.

1   a $7(v + 3) = 5(v + 7)$ b $2(2d + 4) = 4(2d - 7)$
    c $3(x - 2) = -2(x + 19)$ d $2(5 - 2k) = 3(1 + 2k)$

2   a $3(m + 1) + 2(m + 2) = 52$ b $7(d + 2) - 2(2d + 1) = 21$
    c $2(6t + 1) - 4(t - 1) = 0$ d $4(2 - b) + 3(3 + 2b) = 33$

3 The time, T seconds, for a person to travel n floors in a lift is given by the formula $T = 10 + 7n$.

    a   How long does it take to travel:
        i   3 floors?   ii   9 floors?
    b   i   Calculate T when $n = 0$.
       ii   What does your answer mean?

4 N people have a meal in a restaurant and decide to share the cost equally. The amount they each pay is given by the formula $1.1 \times \dfrac{F + D}{N}$

where F is the cost of the food and D is the cost of the drink.
    a   Find how much each person pays if:
        i   5 people visit a pizza parlour where the food costs £20 and the drinks cost £5.
       ii   A couple visits an expensive restaurant where the food costs £44 and the drink costs £24.
    b   What does the 1.1 represent?

5 Show that $\dfrac{a + b}{c} = \dfrac{a}{c} + \dfrac{b}{c}$ is an identity by substituting into both sides:

    a $a = 9$, $b = 5$ and $c = 2$ b $a = 9x$, $b = 12x$ and $c = 3$

6 Show that $(a - b)^2 \equiv a^2 - 2ab + b^2$ is an identity by substituting into both sides:

    a $a = 9$ and $b = 5$ b $a = 5x$ and $b = 2x$

Practice — 3B Simultaneous equations

Solve these simultaneous equations.

1 Subtract the equations.

    a $3x + 2y = 12$
           $3x + y = 9$

    b $5x + 2y = 12$
           $3x + 2y = 8$

    c $4x - y = 14$
           $x - y = 2$

    d $2x + y = 9$
           $2x + 3y = 19$

2 Add the equations.

    a $5x - 2y = 1$
           $4x + 2y = 8$

    b $x + 6y = 9$
           $5x - 6y = 9$

3 Add or subtract the equations.

    a $3x + 5y = 17$
           $3x + 2y = 14$

    b $7x - 2y = 1$
           $4x + 2y = 32$

    c $2x + 3y = 25$
           $5x - 3y = 10$

    d $12x - 2y = 14$
           $7x - 2y = 4$

Practice — 3C Solving by substitution

Solve these simultaneous equations. Rearrange the simpler equation to make x or y the subject. Substitute x or y in the other equation.

1 $x - y = 5$
     $3x - 2y = 16$

2 $5x - 2y = 13$
     $x - 2y = 1$

3 $x + 2y = 12$
     $2x + y = 9$

4 $3x + y = 11$
     $5x + 3y = 21$

5 $5x + 2y = 25$
     $2x + y = 12$

6 $x + y = 13$
     $5x + 3y = 55$

Practice

3D Equations involving fractions

Solve these equations. Show your working.

1
a $\dfrac{x}{4} = 9$
b $\dfrac{w}{7} = -2$
c $\dfrac{m}{0.4} = 0.3$
d $\dfrac{3d}{5} = 6$
e $\dfrac{4u}{3} = 3$
f $\dfrac{2p}{7} = 4$
g $-\dfrac{3m}{5} = 9$
h $\dfrac{2t}{-3} = -5$

2
a $\dfrac{s}{8} = \dfrac{3}{4}$
b $\dfrac{m}{5} = \dfrac{1}{2}$
c $\dfrac{h}{3} = \dfrac{4}{5}$
d $\dfrac{3m}{10} = \dfrac{12}{5}$
e $\dfrac{2j}{3} = \dfrac{6}{7}$
f $\dfrac{3t}{2} = \dfrac{5}{6}$
g $\dfrac{4y}{5} = \dfrac{1}{20}$
h $\dfrac{2n}{7} = -\dfrac{3}{5}$

3
a $h + \dfrac{3}{4} = \dfrac{5}{6}$
b $\dfrac{3}{4}p - 2 = 7$
c $\dfrac{2}{3}a + \dfrac{1}{3} = 1$
d $\dfrac{4}{5}g - \dfrac{3}{5} = \dfrac{3}{5}$
e $\dfrac{1}{2}x + \dfrac{3}{8} = 5$
f $\dfrac{3}{10}c - \dfrac{2}{5} = \dfrac{1}{2}$

4
a $\dfrac{1}{4}b + \dfrac{1}{2}b = 6$
b $\dfrac{4}{5}d - \dfrac{1}{5}d = \dfrac{7}{10}$
c $\dfrac{1}{3}f + \dfrac{5}{6}f = 14$

5
a $\dfrac{x+2}{3} = 2$
b $\dfrac{y-6}{5} = 1$
c $\dfrac{2m+3}{3} = 5$

6
a $\dfrac{r+2}{3} = \dfrac{r+3}{2}$
b $\dfrac{s-3}{2} = \dfrac{2s+1}{5}$
c $\dfrac{4t-3}{6} = \dfrac{t-1}{4}$

7
a $\dfrac{1}{h} = 4$
b $\dfrac{10}{y} = 2$
c $\dfrac{2}{w} = 7$
d $\dfrac{4}{p} = -3$
e $\dfrac{1}{2e} = 1$
f $\dfrac{5}{3i} = 10$
g $\dfrac{8}{3k} = \dfrac{5}{6}$
h $\dfrac{4}{5x} = \dfrac{8}{15}$

8
a $\dfrac{8}{x+2} = \dfrac{4}{x-1}$
b $\dfrac{8}{x} = \dfrac{12}{2x-1}$
c $\dfrac{6}{4x-3} = \dfrac{5}{x+1}$

Practice

3E Inequalities

1 Solve these inequalities. Show your solutions using a number line.

a $5x < 20$
b $y - 5 \geqslant 8$
c $\dfrac{t}{3} \leqslant 2$
d $P + 6 > 15$
e $4c + 9 \geqslant 33$
f $2m - 1 < 27$
g $\dfrac{n}{4} - 2 \leqslant 3$
h $3(x + 5) > 21$
i $2(3e - 2) \geqslant 14$
j $5x + 2 < 2x + 14$

2 Write down the values of x that satisfy these conditions.

a $x - 3 < 4$ where x is a positive integer

b $2x \geqslant 12$ where x is a positive integer less than 10

c $\dfrac{x}{5} \leqslant 2$ where x is a positive even number

d $2x - 5 < 19$ where x is a triangular number

3 Illustrate the solutions to each pair of inequalities using a number line.

   a $x < 10$ and $x \geq 3$
   b $x \leq 9$ and $x > 4$
   c $x < 3$ or $x > 6$
   d $x \leq -5$ or $x \geq 5$

4 Solve these inequalities. Show your solutions using a number line.

   a $2p < 28$ and $p - 7 > 2$
   b $5t - 2 \geq 33$ and $\frac{t}{4} \leq 3$

5 Solve these inequalities. Show your solutions using a number line. Don't forget the negative part of the solution.

   a $x^2 < 36$
   b $x^2 \leq \frac{1}{4}$
   c $x^2 > 100$

Practice

3F Graphs showing direct proportion

1 These tables show the amount charged for different weights of cheddar cheese in two shops.

Country Fare

Weight, W g	200	350	500	600	900
Cost, C pence	84	147	210	243	369

Farm Fresh

Weight, W g	150	400	550	700	1200
Cost, C pence	66	176	242	308	528

   a Calculate the ratio $\frac{Cost}{Weight}$ for each pair of values.
   b For which shop is the cost directly proportional to the weight of cheese? Give a reason for your answer.
   c Write an equation connecting C and W for this shop.
   d **i** How much would this shop charge for 1.6 kg of cheese?
   ii What weight of cheese can be bought for £5?

2 The diagram shows how two businesses charge for the hire of a canoe.

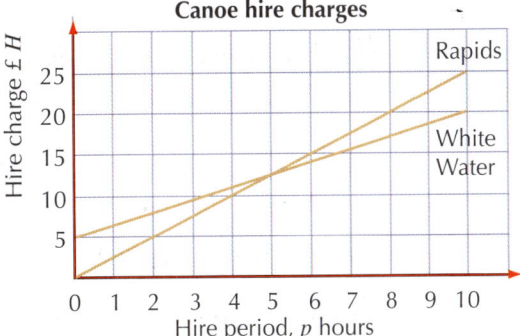

   a For which business is the charge directly proportional to the hire period? Give a reason for your answer.
   b For this business, what is the hire charge for 1 hour?
   c Write an equation connecting the hire charge, H, with the hire period, p.
   d **i** What would this company charge to hire the canoe for 4.5 hours?
   ii What is the hire period corresponding to a hire charge of £9.50? Give your answer in hours and minutes.

3 Chantal measured the distance a wind-up car travelled after different numbers of winds. This table shows her results.

Number of winds, n	5	10	15	20	30	50
Distance travelled, d metres	3.1	5.7	9.3	11.5	18.7	28.7

  a Calculate the ratio $\frac{d}{n}$ for each pair of values.
  b Is the distance travelled roughly directly proportional to the number of winds? Give a reason for your answer.
  c Write an equation connecting d and n.
  d i Estimate how far the car would travel on 25 winds.
     ii Estimate the number of winds needed to make the car travel 24 m.

4 The number of bricks, n, to build d metres of wall is given by the ratio

$n : d = 875 : 25$

  a Simplify the ratio $n : d$.
  b Write an equation connecting n and d.
  c Copy and complete the table.

Length of wall, d metres	5	10	15	20	25
Number of bricks, n					875

  d Calculate the number of bricks needed to build 18 m of the wall.
  e Draw a graph connecting d and n.

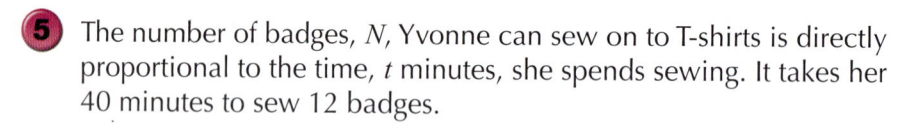

5 The number of badges, N, Yvonne can sew on to T-shirts is directly proportional to the time, t minutes, she spends sewing. It takes her 40 minutes to sew 12 badges.

  a Copy, complete and simplify the ratios.
     i $\frac{\text{Number of badges}}{\text{Time taken}} =$
     ii $N : t =$
  b Write an equation connecting N and t.
  c i How many badges can Yvonne sew in 70 minutes?
     ii How long does it take Yvonne to sew 39 badges?

Practice

3G Solving simultaneous equations by graphs

Solve these simultaneous equations by drawing graphs.

1 Number both axes from 0 to 10.

  a $y + x = 7$
     $y - 2x = 1$
  b $2y - x = 6$
     $y + 2x = 8$
  c $y - 2x = 3$
     $y - 3x = 1$
  d $2y - x = 10$
     $y + x = 8$

2 Number both axes from 0 to 5. Some of the answers are decimals.

  a $y + x = 4$
     $4y - 3x = 4$
  b $5y + 3x = 15$
     $4y - x = 4$

3. Number both axes from –5 to 5. Plot these three graphs on the *same* axes.

   $x + y + 4 = 0$ $x - y + 2 = 0$ $4x - y - 4 = 0$

   Use the graphs to solve each pair of simultaneous equations.

   a $x + y + 4 = 0$
      $x - y + 2 = 0$

   b $x - y + 2 = 0$
      $4x - y - 4 = 0$

   c $4x - y - 4 = 0$
      $x + y + 4 = 0$

CHAPTER 4 Geometry and Measures 1

Practice 4A Pythagoras' theorem

Give all your answers correct to one decimal place.

1. Calculate the length of the hypotenuse.

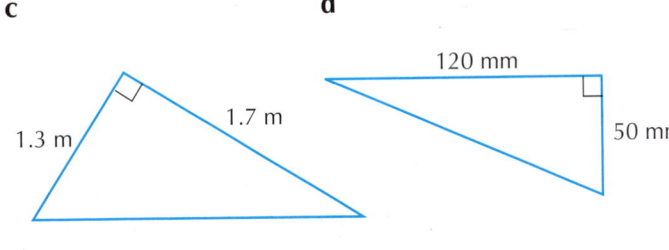

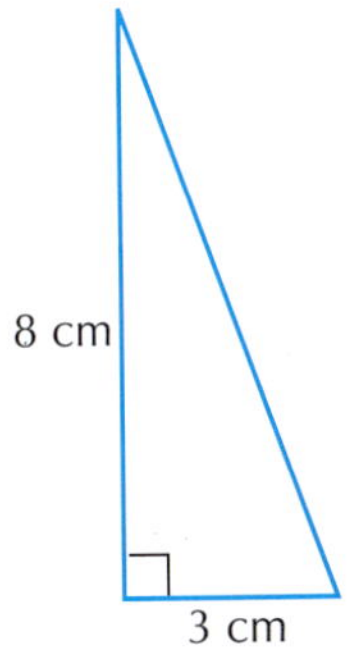

2. Calculate the length of the unknown side.

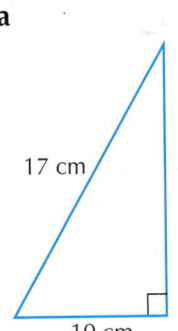
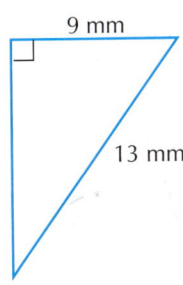
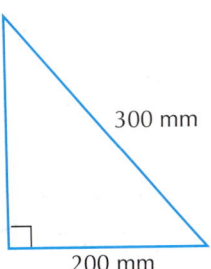
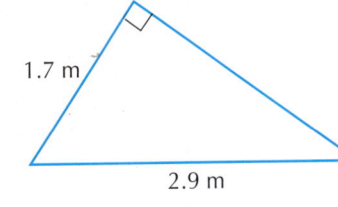

3 a Draw three different right-angled triangles using squared paper.
b Calculate the length of the hypotenuse in each triangle.

4 Calculate the marked lengths.

a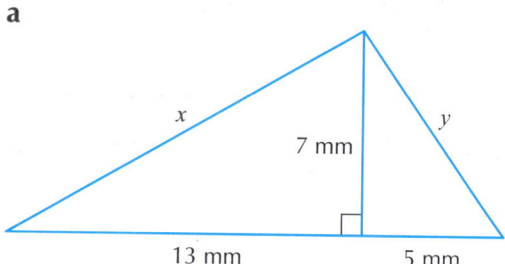

b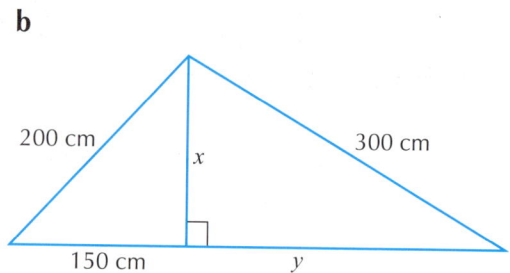

Practice
4B Solving problems using Pythagoras' theorem

Give all your answers correct to a suitable degree of accuracy.

1 Calculate the marked lengths.

a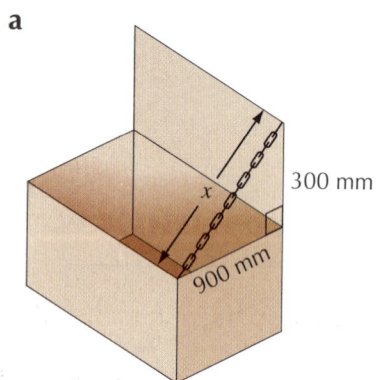

b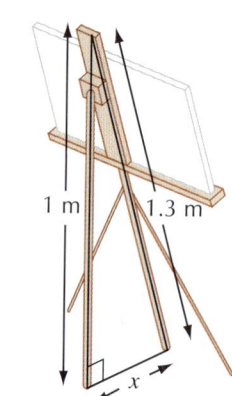

c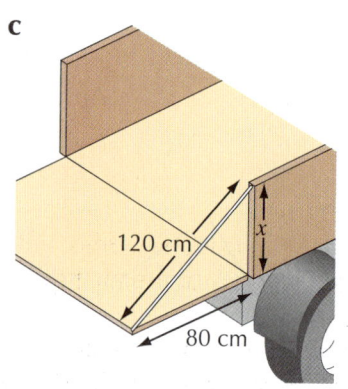

d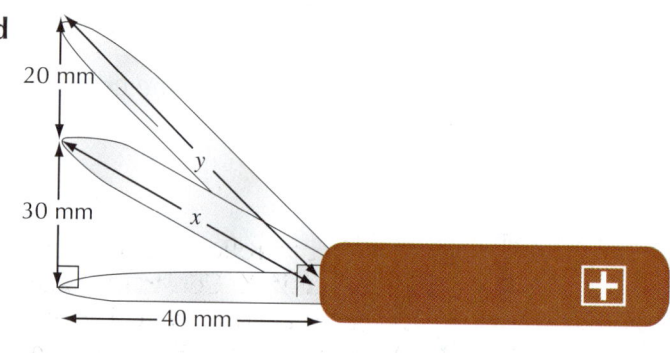

2 The diagram shows the **same** folding ladder in different positions. Calculate the marked lengths.

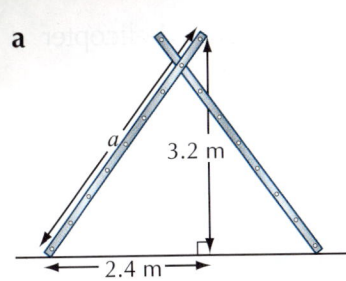

3. Calculate the sides of each of the triangles to the nearest 10 m. Each square represents 100 m.

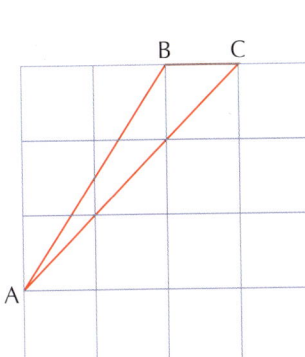
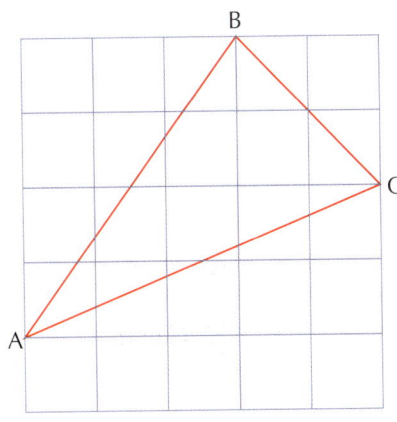

Practice

4C Loci

1. Draw a line AB that is 9 cm long. Use a ruler and compasses only to construct the perpendicular bisector of AB. Mark C as the midpoint of AB.

2. Use a ruler and compasses only to construct the angle bisector of this angle.

 3. a. A radio controlled airship, A, has a range of 400 m from the radio control unit, X. Construct a scale drawing of the locus of the airship using a scale of 1 cm to 100 m.
b. A second airship, B, has a range of less than 300 m from its radio control unit Y. The radio controllers X and Y are 500 m apart. Shade the region where the two airships could collide.
c. Airship, A, steers a course equidistant from the two radio controllers. Draw its locus.

4 This diagram shows a cross-section of a valley APB. There is a helicopter landing pad at P.

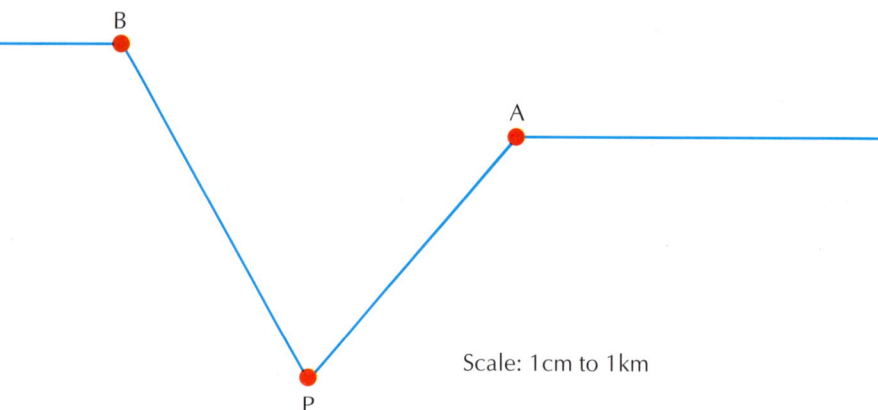

Scale: 1cm to 1km

a Trace the diagram.
b A helicopter, H, flies toward the landing pad, keeping equidistant from beacons A and B. Construct its locus using a dotted line.
c A straight laser tracking beam is continuously transmitted from the landing pad, equidistant from the walls of the valley. Construct its locus using a dotted line.
d When the helicopter reaches the beam, it changes course and flies straight to the landing pad. Use a solid line to draw the flight path of the helicopter.
e Mark the two points, X, where the helicopter is 4 km from beacon A.

Practice 4D Congruent triangles

Give reasons for your answers. State which condition of congruency you are using, i.e. SSS, SAS, ASA or RHS.

1 Show that these pairs of triangles are congruent.

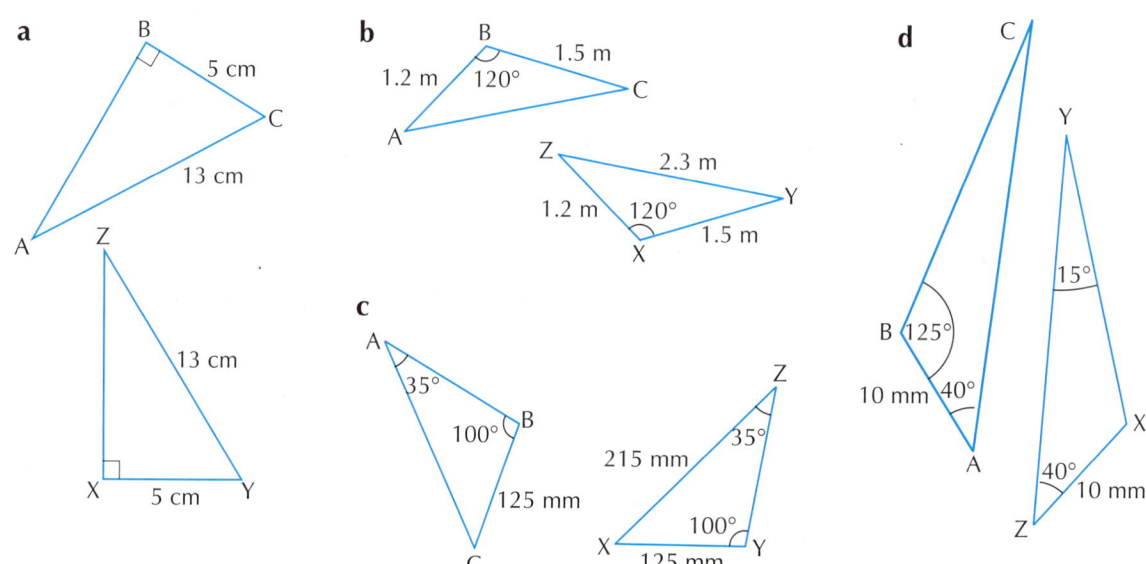

2 Which of these pairs of triangles are congruent?

a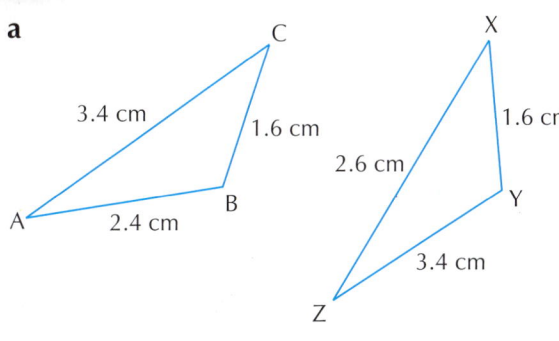

b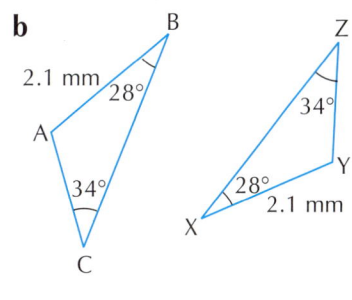

c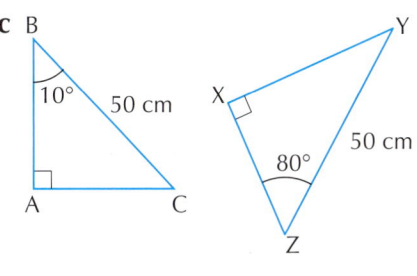

d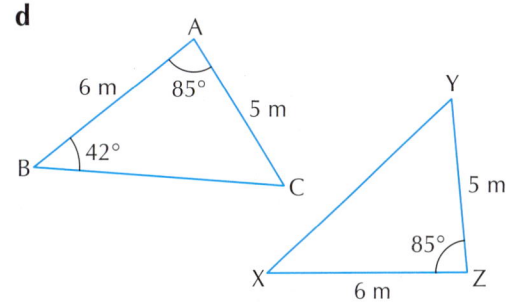

3 a Explain why these triangles are not necessarily congruent.

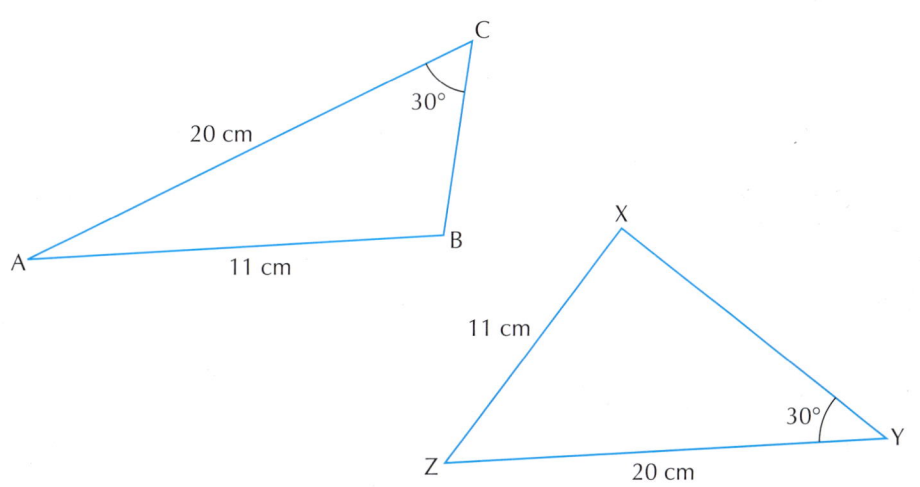

b Check your answer by trying to draw one of the triangles.

4 Draw a kite and label its vertices ABCD. Draw the diagonals and label their intersection E. State all pairs of congruent triangles.

Practice
4E Circle theorems

1 Calculate the marked angles.

a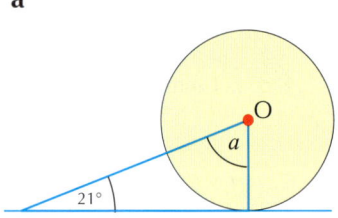

b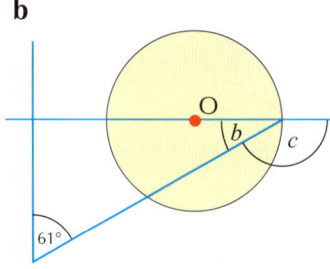

c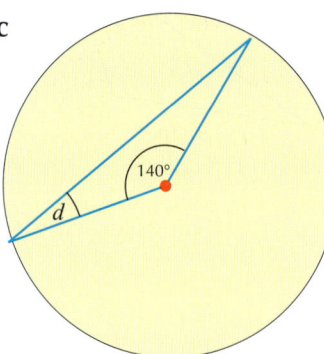

d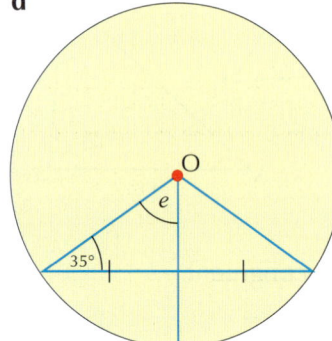

e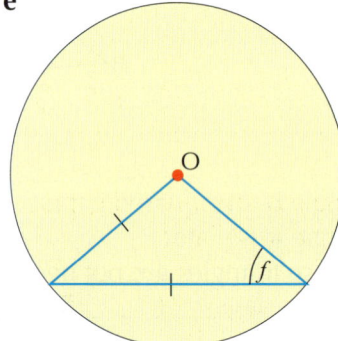

f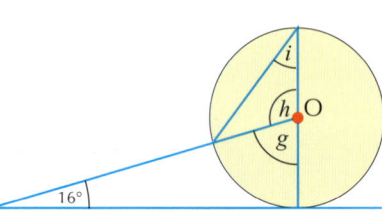

2 Use Pythagoras' theorem to find the marked lengths, correct to one decimal place.

a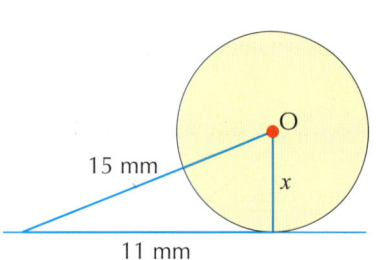

b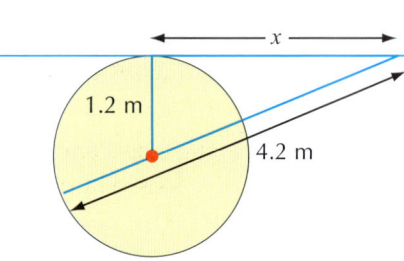

3 Trace this diagram. Construct a circle with centre O and tangent AB. Do not measure any lengths.

4. Trace this diagram. Draw a circle that passes through points A, B and C.

Do not measure any lengths.
Hint: AC and BC will be chords.

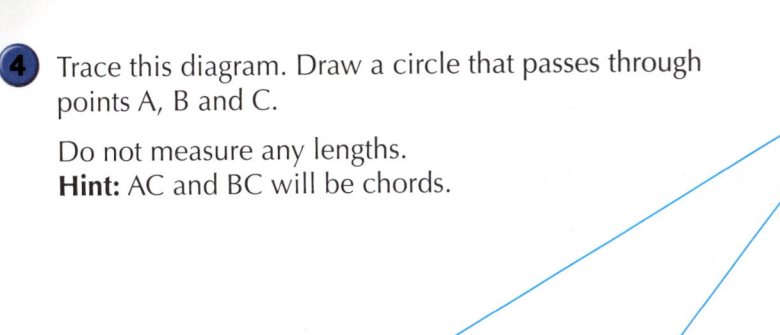

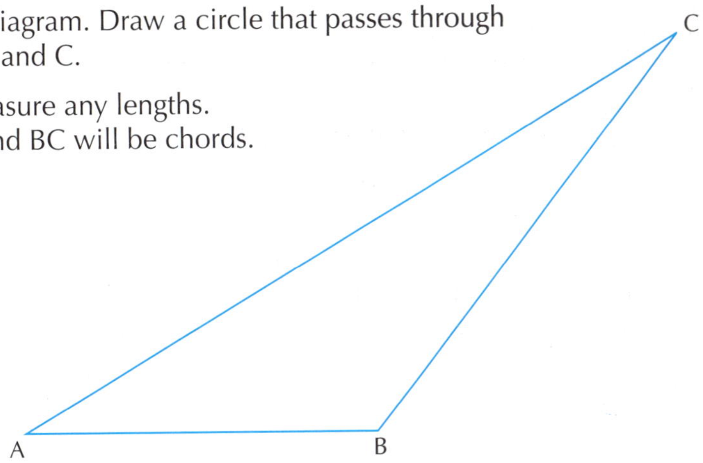

Practice

4F Tessellations

1. **a** Copy each triangle on to squared paper.
   Does each triangle tessellate?

   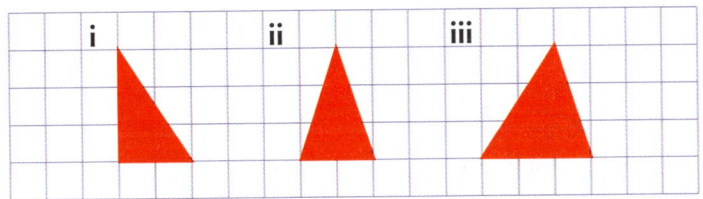

   b Draw a scalene triangle (three unequal sides) of your own on squared paper. Does it tessellate?
   c Does a triangle always tessellate? Explain your answer.

2. **a** Copy each quadrilateral on to squared paper. Does each quadrilateral tessellate?

   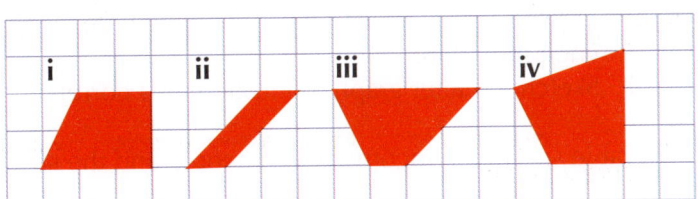

   b Draw a quadrilateral of your own on squared paper (not a special quadrilateral). Does it tessellate?
   c Does a quadrilateral always tessellate? Explain your answer.

CHAPTER 5 Statistics 1

Practice 5A Scatter graphs and correlation

1 There are three ways of combining the variables in pairs (xy, yz and zx). Write down what sort of correlation each pair will have.

  x: The number of people who travel to work by car each month
  y: The number of people who travel to work by public transport each month
  z: The amount of car exhaust pollution each month

Illustrate each correlation by sketching a scatter graph.

2 Ten identical houses had their lofts converted. These scatter graphs show the number of workers, time taken and cost of equipment hire for each loft.

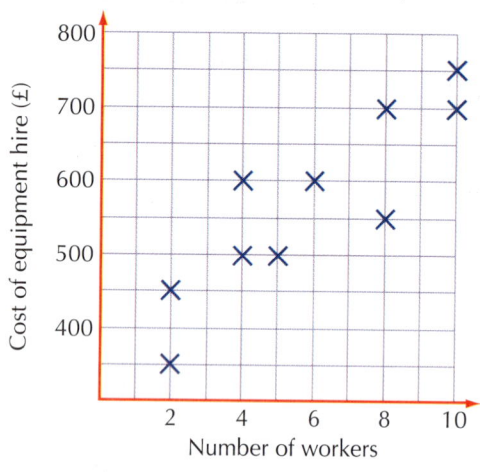
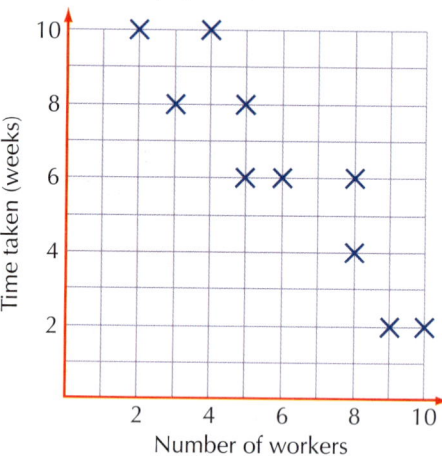

  a Describe the correlation between the number of workers and time taken.
  b Describe the correlation between the number of workers and the cost of equipment hire.
  c i Describe the correlation between the time taken and the cost of equipment hire.
   ii Sketch a scatter graph to illustrate this correlation.

3 The table shows the incomes of 10 people and the size of their car engines.

Income, £	10 000	14 000	17 000	27 000	30 000	36 000	37 000	42 000	45 000	50 000
Engine size, cc	1200	1800	1600	1700	2400	3200	1600	2700	3700	2800

a Draw a scatter diagram using these axes.
  Horizontal axis (income): £10 000 to £50 000
  Vertical axis (engine size): 1000 cc to 4000 cc
b Describe the correlation between income and engine size.

4 Ten pupils spent between 60 and 90 minutes writing a poem. Each poem was marked out of 100. What correlation would you expect between the time spent writing and the marks obtained?

Practice
5B Scatter graphs and lines of best fit

1 This table shows the size and price of books sold by a bookshop in an hour.

Number of pages, n	60	60	120	140	140	180	240	280	320	400	480
Price of book, £P	1.50	1.00	2.50	3.00	3.20	5.00	4.50	6.00	6.00	7.50	10.00

a Draw a scatter graph for the data. Use these scales.
  Number of pages: 2 cm to 100 pages
  Price of book: 1 cm to £1
b Draw a line of best fit, by eye.
c Use your line of best fit to estimate:
  i the cost of a book containing 260 pages.
  ii the number of pages in a book costing £1.50.

2 Fifteen young drivers were caught speeding through a village. Their ages and the speeds are shown in the table below.

Age	17	17	18	19	20	20	21	22	22	23	24	25	27	27	29
Speed (mph)	46	54	50	44	48	50	42	42	46	46	42	46	40	36	40

a Draw a scatter graph for the data.
b Use your graph to estimate:
  i the age of a driver caught speeding at 45 mph.
  ii the speed of a driver aged 26.

Practice
5C Time series graphs

1 This diagram shows how the US produced its energy during the last century.

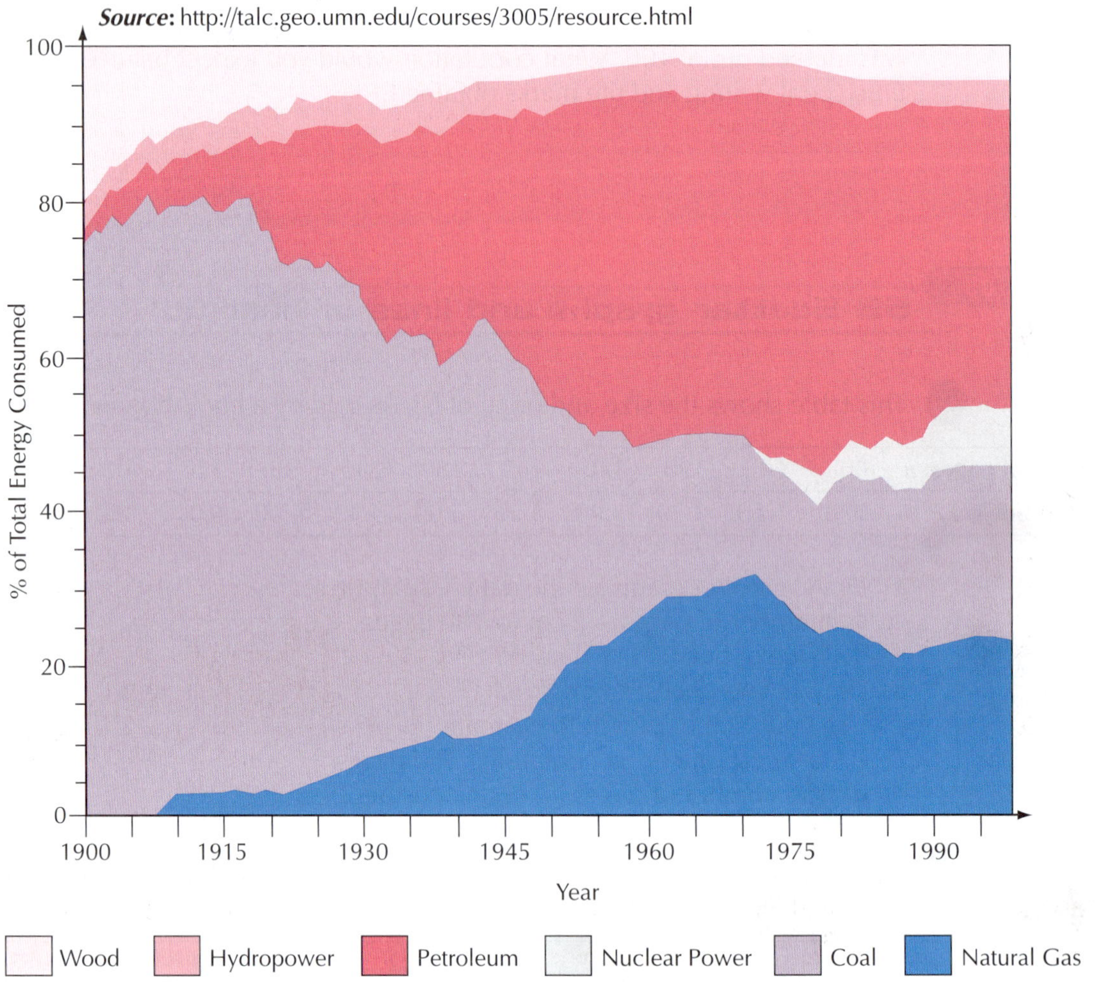

Source: http://talc.geo.umn.edu/courses/3005/resource.html

Legend: Wood, Hydropower, Petroleum, Nuclear Power, Coal, Natural Gas

The height of each stripe is the percentage of the total energy that the power source provides. For example, in 1900, hydropower provided about 4% of the total energy in the US.

a What was the main source of energy at the:
  i beginning of the century?
  ii end of the century?
b Which source of energy provided roughly the same percentage of energy throughout the century?
c Estimate when petroleum and coal were used in equal amounts.
d What percentage of US energy was produced from natural gas in 1975?
e What percentage of US energy was produced from coal in 1945?
f Describe the trend of using wood as an energy source.
g Which energy source provided about twice the energy as hydropower at the end of the century?
h Describe the trend in the use of natural gas.
i The total energy consumed in 1990 was 82 quadrillion units.
  a Find out what a quadrillion is.
  b Estimate the energy produced from petroleum in 1990.

 2 These graphs show the prices and sales of shares in two companies, James, James and Yates (JJY) and Standard Wire (SW), during a day of trading. The number of shares sold is called the 'volume'.

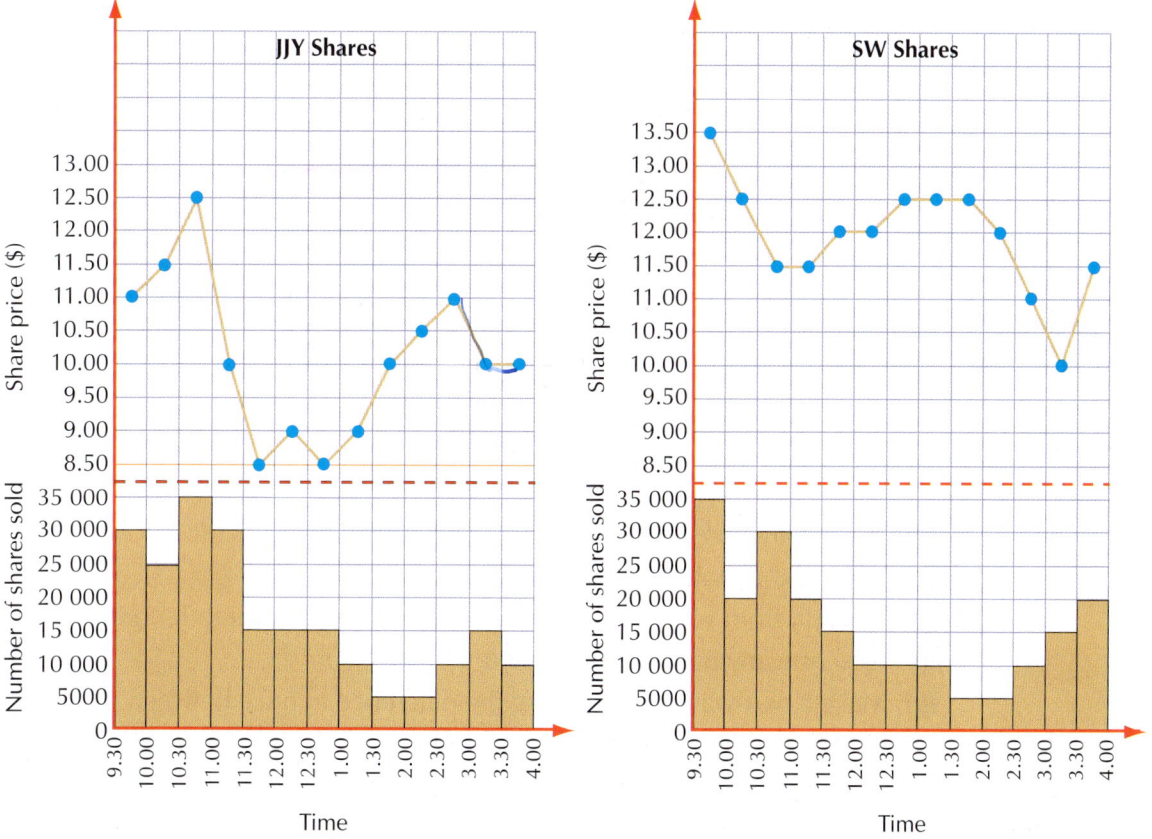

The prices shown are at the end of each 30-minute period. For example, the price of SW was $11.50 at 4 pm, and 20 000 shares were sold between 3.30 pm and 4 pm.

a Which share started trading at the highest price of the day?
b Which share has the greatest range of prices?
c What is the greatest price increase during a half-hour period?
State the share and when it occurred.
d Bad news caused one of the shares to greatly reduce in price.
  i Which share was this?
  ii When did this happen?
  iii What was the fall in price?
e What was the difference in share prices at 12.30 pm?
f What do you notice about the volume during the day for both shares? Give a reason why this might have happened.
g During which hour was the least number of shares traded for each company?
h How much money was spent on JJY shares from 2.30 pm to 3.00 pm?
i During which half-hour period was most money spent altogether? How much was spent?

Practice
5D Two way tables

 1 The tables show some holiday prices (£) to Tenerife and Gran Canaria.

Date	Tenerife 1 week S/C	H/B	A/I	2 weeks S/C	H/B	A/I	Gran Canaria 1 week S/C	H/B	A/I	2 weeks S/C	H/B	A/I
28 Feb	159	269	335	200	320	375	177	309	319	225	349	350
10 Mar	159	275	339	200	325	377	189	309	319	220	349	359
14 Mar	170	285	349	208	329	375	190	315	335	218	340	339
21 Mar	175	289	349	215	340	380	190	319	335	220	340	339
28 Mar	149	279	339	215	320	380	195	330	339	216	365	345
4 Apr	169	269	345	219	315	375	199	389	379	225	419	399

S/C = self-catering; HB = hotel, half-board; A/I = all-inclusive

a What is the most expensive two-week holiday in March?
b What is the range of prices of one-week holidays?
c Overall, from which date is it most expensive to travel to Tenerife?
d Describe any patterns you notice.
e Make a two-way table showing the difference in price between one-week and two-week holidays to Gran Canaria. What do you notice?
f Which type of holiday to Tenerife (S/C, HB or A/I) varies in price the least? Explain your answer.

 2 The table shows the percentage of people owning a computer in 1999 and 2005, and the percentage of the population running a computer connected to the Internet.

		1999 Owns a computer connected to the Internet (%)	Owns a computer (%)	2005 Owns a computer connected to the Internet (%)	Owns a computer (%)
Age	15–24	13	45	58	73
	25–34	14	47	61	72
	35–44	17	52	70	79
	45–54	15	48	63	71
	55–64	10	30	41	50
	65–74	4	15	14	21
	75+	2	7	8	11

a Describe any trends you notice.
b Compare 1999 with 2005.
c Construct a two-way table showing the percentages of people owning computers not connected to the Internet.

Practice

5E Cumulative frequency diagrams

1 This table shows the volume of ice slush dispensed by a number of machines in a chain of shops.

Volume of ice slush (v litres)	Number of machines
$0 \leqslant v < 2$	3
$2 \leqslant v < 4$	8
$4 \leqslant v < 6$	12
$6 \leqslant v < 8$	9
$8 \leqslant v < 10$	5
$10 \leqslant v < 12$	2

a Copy and complete the cumulative frequency table below.

Volume of ice slush (v litres)	Number of machines
$v < 0$	0
$v < 2$	3

b Draw a cumulative frequency diagram.
c Use your graph to estimate the number of machines dispensing:
  i less than 5 litres of slush.
  ii less than 9 litres of slush.
  iii between 5 and 9 litres of slush.
  iv more than 3 litres of slush.
d Use your graph to estimate the:
  i median.
  ii lower quartile.
  iii upper quartile.
  iv interquartile range.

2 The areas of photographs used in a catalogue are shown below.

Area of photograph (A cm²)	Number of photographs
$0 < A \leqslant 5$	13
$5 < A \leqslant 10$	25
$10 < A \leqslant 20$	52
$20 < A \leqslant 30$	61
$30 < A \leqslant 40$	34
$40 < A \leqslant 60$	11

a Construct a cumulative frequency table.
b Draw a cumulative frequency diagram.

c Use your graph to estimate the number of photos of area:
  i   less than 15 cm².
  ii less than 35 cm².
  iii between 15 and 35 cm².
  iv greater than 25 cm².
d Use your graph to estimate the:
  i   median.
  ii lower quartile.
  iii upper quartile.
  iv interquartile range.

Practice

5F Estimation of a mean from grouped data

1 This table shows the distances a discus was thrown in a competition.

Distance, d metres	Number of competitors, f	Mid-class value, x	$f \times x$
$15 \leqslant d < 20$	7		
$20 \leqslant d < 25$	20		
$25 \leqslant d < 30$	22		
$30 \leqslant d < 35$	7		
$35 \leqslant d < 40$	3		
$40 \leqslant d < 45$	1		
	Total =		Total =

a Copy and complete the table.
b Estimate the mean distance the discus was thrown.

2 A newspaper article claims that the average person spends over 8 hours exercising per week. Matt interviewed 50 people and estimated their weekly amount of exercise. This table shows his results.

Weekly exercise (E hours)	Number of people
$0 \leqslant E < 1$	11
$1 \leqslant E < 3$	5
$3 \leqslant E < 5$	8
$6 \leqslant E < 8$	14
$8 \leqslant E < 10$	6
$10 \leqslant E < 12$	4
$12 \leqslant E < 16$	2

a Copy the table. Add columns as in question 1.
b Estimate the mean amount of exercise per week.
c Does Matt's data support the newspaper article? Explain your answer.

3 These tables show the amounts collected for charity by pupils of Year 8 and Year 9.

Year 8	
Amount collected, £m	Number of pupils
$0 \leqslant m < 5$	15
$5 \leqslant m < 10$	20
$10 \leqslant m < 15$	16
$15 \leqslant m < 25$	12
$25 \leqslant m < 40$	6

Year 9	
Amount collected, £m	Number of pupils
$0 \leqslant m < 10$	11
$10 \leqslant m < 20$	14
$20 \leqslant m < 30$	10
$30 \leqslant m < 40$	7
$40 \leqslant m < 50$	1

a Copy both tables. Add columns as in question 1.
b Which Year collected the most money?
c Estimate the mean amount collected for each Year.
d Did pupils in Year 9 collect more, on average? Explain your answer.

Practice 5G Statistical investigations

Conduct an investigation into one of these areas. Follow the steps in section 5G of Pupil Book 3.

1 Typing

2 Computers in the home

3 Boiling an egg

Chapter 6: Geometry and Measures 2

Practice 6A Similar triangles

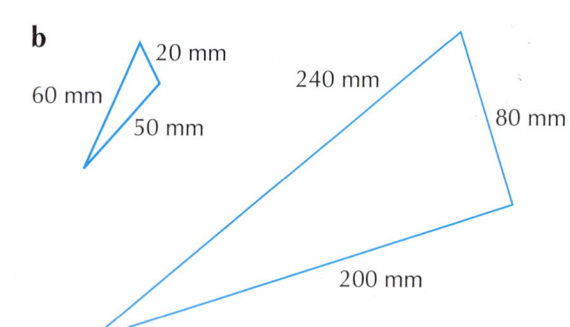

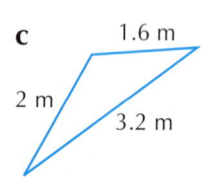

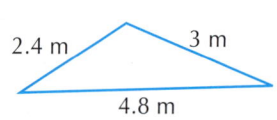

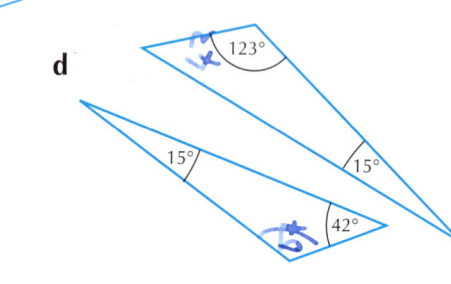

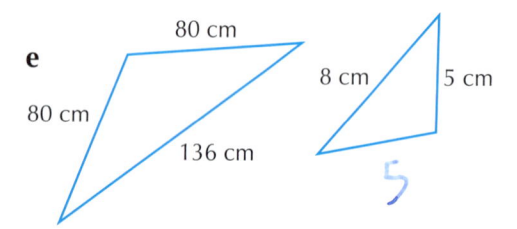

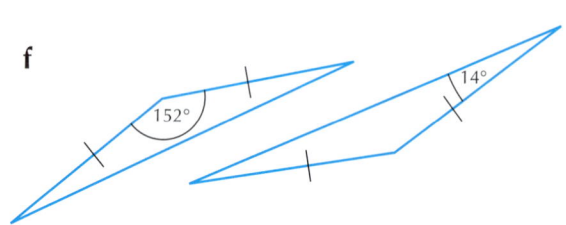

1 State whether the triangles are similar. Give reasons for your answers.

a Triangle 1: angles 50°, 84°. Triangle 2: angles 84°, 36°.

b Triangle 1: sides 20 mm, 60 mm, 50 mm. Triangle 2: sides 240 mm, 80 mm, 200 mm.

c Triangle 1: sides 2 m, 1.6 m, 3.2 m. Triangle 2: sides 2.4 m, 3 m, 4.8 m.

d Triangle 1: angles 123°, 15°. Triangle 2: angles 15°, 42°.

e Triangle 1: sides 80 cm, 80 cm, 136 cm. Triangle 2: sides 8 cm, 5 cm.

f Triangle 1: angle 152°, two equal sides marked. Triangle 2: angle 14°, two equal sides marked.

2 a Explain why triangle ABC is similar to triangle PQR.
b Calculate the unknown lengths.

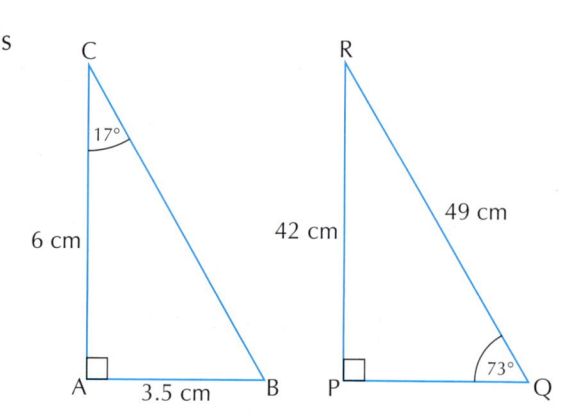

Triangle ABC: right angle at A, angle C = 17°, AC = 6 cm, AB = 3.5 cm.
Triangle PQR: right angle at P, angle Q = 73°, QR = 49 cm, PR = 42 cm.

3 Triangle ABC is similar to triangle PQR.
Calculate the unknown lengths.

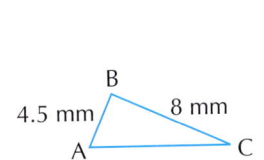

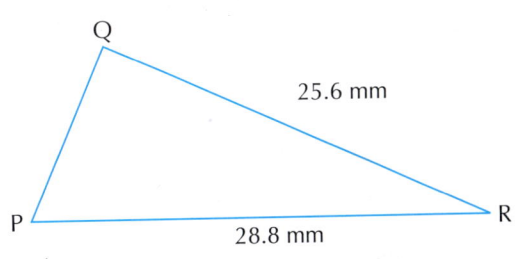

4 The diagram shows the sail of a yacht.
  a Explain why triangles PST and PQR are similar.
  b Calculate the length of the spar ST.

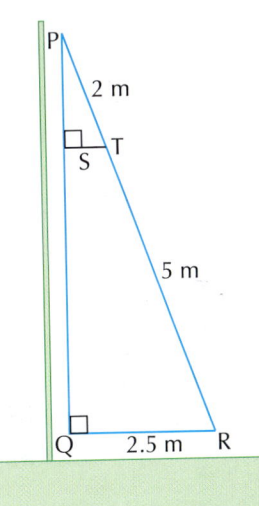

5 a Explain why triangles ABC and CDE are similar.
  b Calculate the length CD.

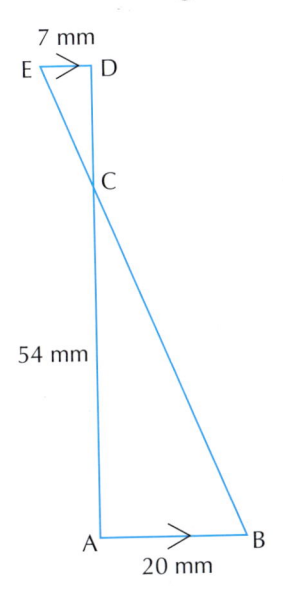

Practice 6B Metric units for area and volume

1 Convert these units.

  a 9 m² to cm²
  b 0.05 m² to cm²
  c 5000 cm² to m²
  d 17 cm² to m²
  e 11 cm² to mm²
  f 850 mm² to cm²
  g 0.065 m² to mm²
  h 28 400 mm² to m²
  i 75 ha to m²
  j 0.085 ha to m²
  k 39 000 m² to ha
  l 1 ha to mm²

2 Convert these units.

  a 9 cm³ to mm³   b 0.066 cm³ to mm³
  c 3 m³ to cm³   d 0.0004 m³ to cm³
  e 650 mm³ to cm³   f 22 500 cm³ to m³
  g 6 m³ to mm³   h 1 400 000 mm³ to m³

3 Convert these units.

  a 72 litres to ml   b 0.15 litres to ml
  c 700 ml to litres   d 27 500 ml to litres
  e 4.2 litres to cm³   f 4000 cm³ to litres
  g 39 cm³ to ml   h 5.2 m³ to litres
  i 80 cm³ to litres   j 4300 ml to cm³
  k 24 000 litres to m³   l 850 litres to m³

4 A rectangular aircraft hangar measures 940 m by 750 m. Find its area, giving your answer in hectares.

5 A rectangular stamp measures 32 mm by 34 mm.

  a What is the area of the stamp in: **i** mm²? **ii** cm²?
  b A sheet of stamps has an area of 1.632 m². How many stamps are there?

6 An inflatable dinghy has a capacity of 0.66 m³. Martha's pump blows 1.2 litres of air into the dinghy every second. How long will it take her to fill the dinghy?

Practice 6C Length of an arc and area of a sector

Use π = 3.142 or π on your calculator.

Give your answers correct to 3 significant figures, where necessary.

1 Calculate: **i** the length of arc and **ii** the area of the sector.

a 10 cm, 45°
b 2.4 m, 100°
c 50 mm, 330°
d 36 cm, 3°

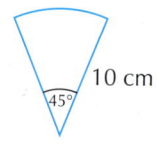

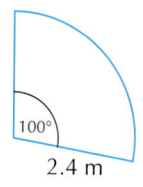

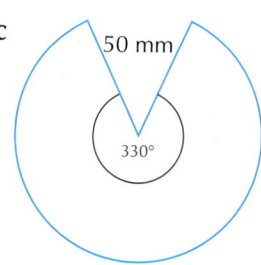

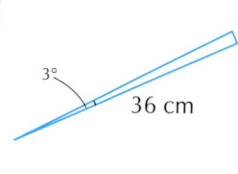

2 Calculate the perimeter of this fan.

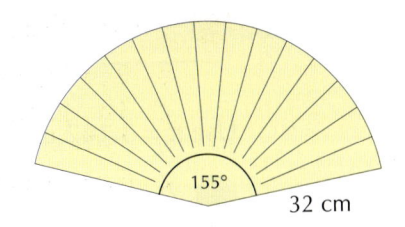

155°
32 cm

3 The diagram shows the cross-section of a lens.

Calculate the perimeter of the cross-section.

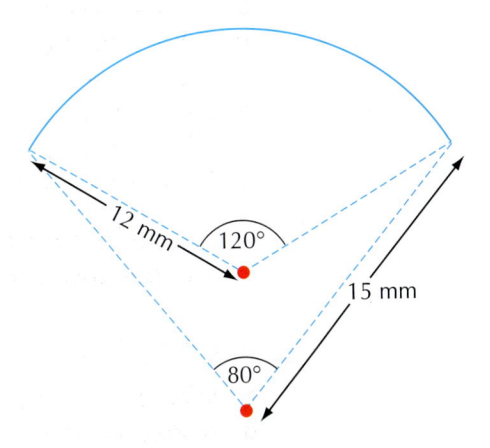

12 mm
120°
15 mm
80°

4 The diagram shows a square corner holder in a photo album.

Calculate the shaded area.

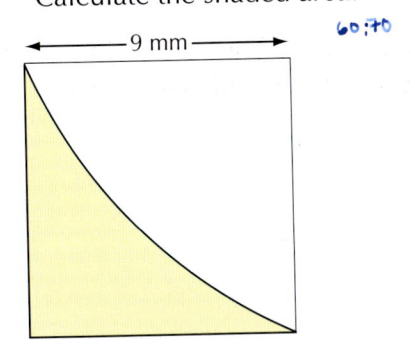

60;70

9 mm

5 Calculate the angle θ to the nearest degree.

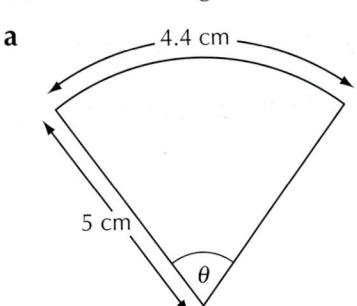

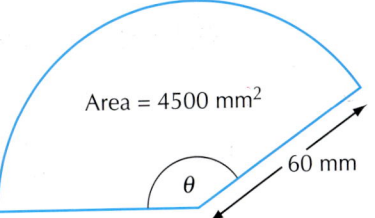

a 4.4 cm
5 cm
θ

b Area = 4500 mm²
θ
60 mm

Practice
6D Volume of a cylinder

Use π = 3.142 or [π] on your calculator.

1 Calculate the volume of each cylinder, correct to the nearest cm³.

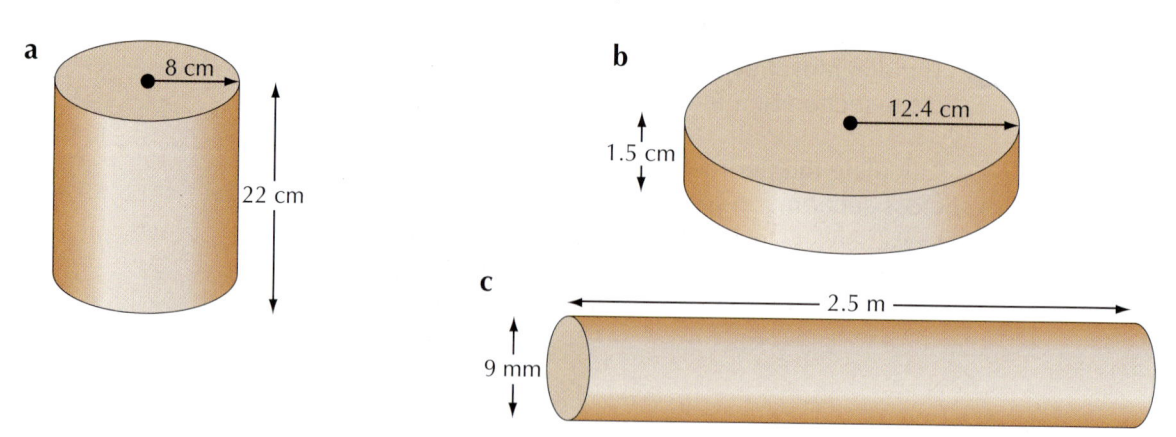

a 8 cm, 22 cm

b 12.4 cm, 1.5 cm

c 2.5 m, 9 mm

2 A company manufactures tubes and tins of shoe polish.

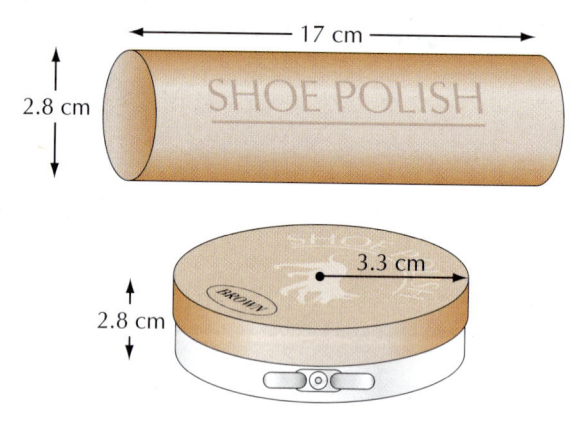

17 cm, 2.8 cm

SHOE POLISH

3.3 cm, 2.8 cm

Which contains the most polish?

3 A cylindrical telegraph pole has a diameter of approximately 21 cm. It is sunk approximately 2.5 m into the ground. What is the volume of the hole?

4 The diagram shows a toilet roll.
  a Calculate the volume of the hole in the middle.
  b Calculate the volume of paper.

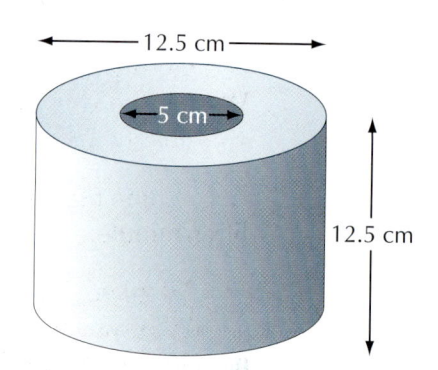

12.5 cm, 5 cm, 12.5 cm

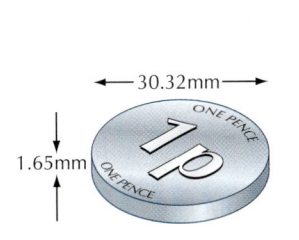

5 1p coins have filled this wishing well to a height of about 25 cm.

- 30.32 mm
- 1.65 mm
- 1 m
- Make a wish for 1p (all proceeds go to charity)

a Calculate the volume of a 1p coin, in cm³.
b Estimate the volume of coins in the wishing well.
c Estimate the number of coins in the wishing well.

Practice
6E Rate of change

1 It took Apollo 11 approximately 73 hours to travel 385 000 km to the Moon. Calculate the average speed of the spaceship.

2 An arrow travelled 130 m at an average speed of 50 m/s. How long was the arrow in the air?

3 A train took $2\frac{1}{2}$ hours to travel from Diston to Hartley at an average speed of 50 mph. The return journey took 90 minutes.

a Calculate the distance from Diston to Hartley.
b Calculate the average speed over both journeys.

4 Two greyhounds, Charlie and Belle, started a race at the same time.

a Charlie completed the race in 3 minutes at an average speed of 50 km/h. How far did he travel?
b Belle had an injury and completed the race in 4 minutes. What was her average speed in: **i** m/s? **ii** km/h?

5 200 cm³ of butter has a mass of 172 g. Calculate the density of butter.

6 The density of oak is 0.7 g/cm³. What is the volume of a plank of oak with mass 3.2 kg?

FM 7 A hosepipe fills this cylindrical can at a rate of litres/minute.

- 30 cm
- 20 cm

a Calculate: **i** the volume of the can in cm³.
ii the capacity of the can in litres.
b How long does it take to fill the can? Give your answer to the nearest second.

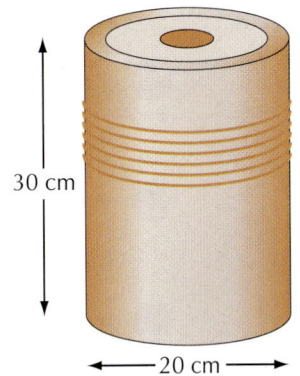

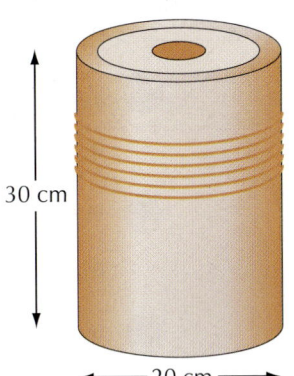

8 On a journey lasting 23 minutes, this bicycle wheel rotated at a rate of 95 revolutions per minute.

   a Calculate the circumference of the wheel.
   b Calculate the length of the journey. Give your answer in km, correct to the nearest 100 m.

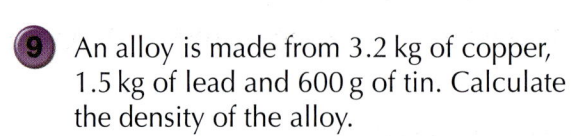

42 cm

9 An alloy is made from 3.2 kg of copper, 1.5 kg of lead and 600 g of tin. Calculate the density of the alloy.

Metal	Density (g/cm³)
Copper	8.96
Lead	11.4
Tin	7.3

CHAPTER 7 Number 2

Practice

7A Standard form

1 Which of these numbers are written in standard form?

   a 40×10^3 **b** 2.3×10^7 **c** 4×10^{-3}
   d 1×10^{10} **e** 10^9 **f** 0.8×10^5

2 Write these as ordinary numbers.

   a 2.1×10^3 **b** 4×10^8 **c** 3.9×10^6 **d** 9.9×10^9

3 Write these as ordinary numbers.

   a 7.9×10^{-2} **b** 4×10^{-7} **c** 4.6×10^{-4} **d** 1.8×10^{-5}

4 Write these as ordinary numbers.

   a 3.7×10^2 **b** 2.8×10^{-3} **c** 8×10^{10} **d** 6×10^{-1}

5 Write these numbers using standard form.

   a 3200   **b** 900 000   **c** 420   **d** 8 000 000 000

6 Write these numbers using standard form.

    a 0.054   **b** 0.000 007   **c** 0.3   **d** 0.0044

7 Write these numbers using standard form.

    a 500   **b** 0.000 67   **c** 0.07   **d** 77 000 000

8 Write these numbers using standard form.

    a 27×10^2 **b** 0.07×10^7 **c** 400×10^{-5} **d** 0.13×10^{-2}

9 Write these quantities in standard form.

    a The distance from the sun to the nearest star: 40 000 000 000 000 km
    b The mass of an electron: 0.000 000 000 000 000 000 000 000 000 000 91 kg
    c The area of the Earth: 510 000 000 km^2

Practice

7B Multiplying with numbers in standard form

1 Calculate these. Give your answers in standard form.
Do not use a calculator. Show your working.

    a $(4 \times 10^3) \times (2 \times 10^2)$ **b** $(7 \times 10^4) \times (4 \times 10^5)$
    c $(8 \times 10^8) \times (1 \times 10^{-3})$ **d** $(2.5 \times 10^{-4}) \times (6 \times 10^{-1})$
    e $(1.2 \times 10^{-7}) \times (7 \times 10^6)$ **f** $(7 \times 10^5)^2$

Use a calculator for the remaining questions.

2 Calculate these. Give your answers in standard form.
Do not round your answers.

    a $(1.4 \times 10^5) \times (2.3 \times 10^3)$ **b** $(3.7 \times 10^{-2}) \times (4.1 \times 10^5)$
    c $(8.7 \times 10^4) \times (3 \times 10^{-6})$ **d** $(4.5 \times 10^{-4}) \times (3.5 \times 10^{-3})$
    e $(9.2 \times 10^{-3}) \times (7.1 \times 10^3)$

3 Calculate these. Give your answers in standard form.
Round your answers to 3 significant figures.

    a $(2.74 \times 10^{-3}) \times (1.9 \times 10^2)$ **b** $(7.33 \times 10^2) \times (4.25 \times 10^8)$
    c $(9.08 \times 10^{-5}) \times (1.1 \times 10^3)$ **d** $(2.59 \times 10^{-4}) \times (6.993 \times 10^{-6})$
    e $(3.29 \times 10^{-3}) \times (7.005 \times 10^{10})$

4 The mass of a proton is 1.67×10^{-27} kg. What is the total mass of 5×10^{18} protons?

5 The distance of the Earth from the Sun is 1.526×10^9 m. A rocket travels from the Earth to the Sun at an average speed of 450 metres per second. How long will the journey take? Give your answer correct to 3 significant figures.

Practice
7C Dividing with numbers in standard form

1 Calculate these. Give your answers in standard form.
Do not use a calculator. Show your working.

  a $(4 \times 10^6) \div (2 \times 10^2)$
  b $(7 \times 10^4) \div (4 \times 10^6)$
  c $(8 \times 10^8) \div (1 \times 10^{-3})$
  d $(2.5 \times 10^7) \div (5 \times 10^3)$
  e $(1.2 \times 10^{-7}) \div (6 \times 10^6)$
  f $(9.99 \times 10^{21}) \div (3 \times 10^{18})$

Use a calculator for the remaining questions.

2 Calculate these. Give your answers in standard form.
Do not round your answers.

  a $(1.4 \times 10^5) \div (2.5 \times 10^3)$
  b $(3.7 \times 10^{-2}) \div (4 \times 10^5)$
  c $(2.4 \times 10^4) \div (7.5 \times 10^{-1})$
  d $(4.5 \times 10^{-4}) \div (7.2 \times 10^{-3})$
  e $(1 \times 10^{-3}) \div (1.25 \times 10^3)$

3 Calculate these. Give your answers in standard form.
Round your answers to 3 significant figures.

  a $(8.73 \times 10^{-3}) \div (1.9 \times 10^2)$
  b $(4.33 \times 10^2) \div (2.25 \times 10^8)$
  c $(1 \times 10^{-5}) \div (1.1 \times 10^3)$
  d $(6.59 \times 10^{-4}) \div (2.993 \times 10^{-6})$
  e $(7.29 \times 10^{-3}) \div (2.005 \times 10^{10})$
  f $\sqrt[3]{3.4 \times 10^{20}}$
  g $\dfrac{1}{1.4 \times 10^4}$

4 A digital photo has a size of 1.2×10^6 bytes. How many photos can be stored on a computer hard drive with a capacity of 2.8×10^{11} bytes?

5 Light travels at a speed of 2.99×10^8 metres per second. The distance of the Earth from the Sun is 1.526×10^6 km. How long does it take for light to travel from the Sun to the Earth?

Practice
7D Upper and lower bounds 1

1 Find the upper and lower bounds of these quantities.

  a The radius of a pizza that is 13 cm, to the nearest cm.
  b A piggy bank that holds 500 1p coins, to the nearest 10 coins.
  c A battery that lasts 60 hours, to the nearest 5 hours.
  d A computer that downloads at a speed of 62 000 bytes per second, to the nearest 1000 bytes per second.
  e A car that can travel 800 miles on a full tank, to the nearest 50 miles.

2 ChocoDrops are covered with 60 sugar dots, to the nearest 10.
Bags contain 120 ChocoDrops, to the nearest 10.

  a What is the least number of sugar dots on a ChocoDrop?
  b What is the greatest number of ChocoDrops in a bag?
  c What is the least number of sugar dots contained in a bag?

9 These two candles are similar.

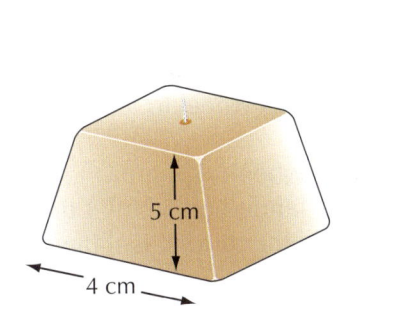

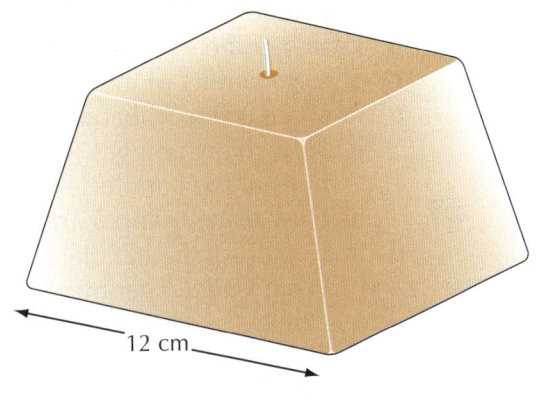

5 cm

4 cm

12 cm

a Calculate the height of the large candle.
b The area of the top of the small candle is 7 cm². Calculate the area of the top of the large candle.
c The volume of the small candle is 56 cm³. Calculate the volume of the large candle.

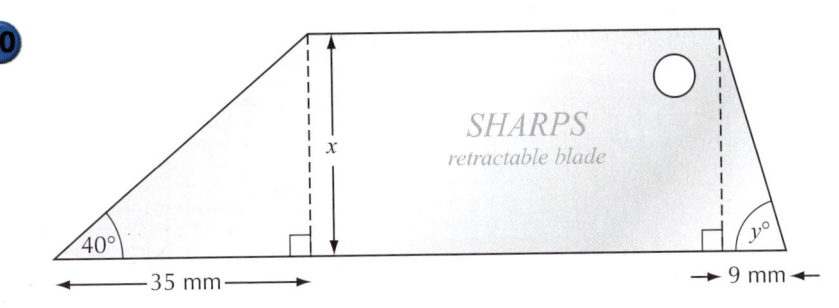

10

x

SHARPS
retractable blade

40° $y°$

35 mm   9 mm

a Calculate the height, x, of the blade.
b Calculate the angle, $y°$, of the handle.

Practice
12F Statistics

You may use a calculator.
Show your working.

1 A fruit machine has three wheels.
Each wheel has 25 symbols.
The table shows the number of each symbol on the first wheel.

Symbols on wheel 1	
Apple	7
Cherry	2
Banana	4
Pear	5
Bell	1
Orange	6

LUCKY SEVEN

a For each pull of the handle, what is the probability that the first wheel will land on:
   i a cherry?   ii a fruit?   iii an apple or orange?
b Which of the events in part **a** are unlikely to occur?
c The probability of the second wheel landing on an orange is 0.2.
   i What is the probability of it not landing on an orange?
   ii How many oranges would you expect this wheel to land on with 150 pulls of the handle?
   iii How many oranges are on the second wheel?
d In 20 pulls of the handle, the third wheel landed on a bell 3 times.
   i What is the experimental probability of the third wheel landing on a bell?
   ii Estimate the number of bells on the wheel.
   iii How could you obtain a better estimate?

2 StarGirl is a handheld electronic game with five levels of difficulty. When a player scores 100 points in one turn, they start the next level.

a Marcus obtained these scores on Level 1.
   20   0   60   40   80   40   40   80   60   0   100
   Calculate these statistics.
   i Mode   ii Median   iii Mean   iv Range
b At the end of Level 1, Marcus received a bonus of 60 points. Explain how this affects each of the statistics in part **a**.
c This table summarises Marcus' scores up until he completed Level 5.

Score	Number of turns
0	12
20	20
40	45
60	38
80	29
100	5

   i What is his modal score?
   ii Calculate the total points he scored.
   iii Calculate his mean score.

3 Hing Wai conducted an investigation into the amount of fiction and non-fiction adults watch on TV. He asked ten of his teachers these questions.

'How many hours do you spend watching films each week?'
'How many hours do you spend watching documentaries each week?'
His report included this table of results and scatter diagram.

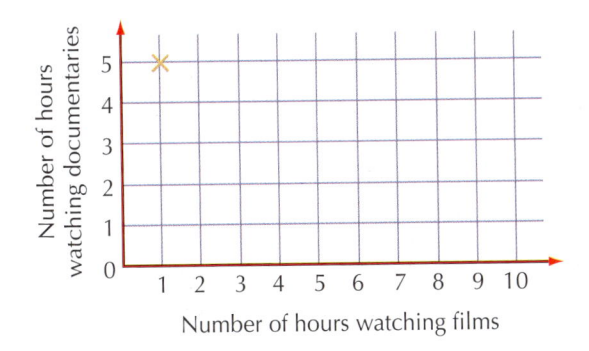

Teacher	Ms Witon	Mr Carter	Mr Singh	Mrs Tenby	Mrs Dean	Mr Chan	Ms Aldridge	Mr Jones	Ms Foster	Mr Phillips
Number of hours watching films	0	1	5	4	3	2	5	3	4	3
Number of hours watching documentaries	4	5	0	2	4	3	3	4	3	2

a Copy and complete the scatter diagram.
b Draw a line of best fit on your diagram.
c Describe any correlation.
d Use your line of best fit to estimate:
  i the number of hours spent watching films when $3\frac{1}{2}$ hours are spent watching documentaries.
  ii the number of hours spent watching documentaries when 9 hours are spent watching films.
e What conclusion could Hing Wai make, based on this data?
f Give a reason why this conclusion may not be true.
g Write down another useful question that Hing Wai could have asked.

4 The table shows the amounts patients paid when they visited a private dentist.

Cost, £C	Number of patients, f	Mid-class value, x	$f \times x$
$0 < C \leq 20$	5		
$20 < C \leq 40$	3		
$40 < C \leq 60$	9		
$60 < C \leq 100$	11		
$100 < C \leq 140$	8		
$140 < C \leq 180$	4		
	Total =		Total =

a Copy and complete the table.
b Estimate the mean cost of visiting the dentist.

5 The table shows the treatment times of patients visiting a dentist.

Time (t minutes)	Number of patients
$10 \leqslant t < 20$	3
$20 \leqslant t < 30$	12
$30 \leqslant t < 40$	18
$40 \leqslant t < 50$	14
$50 \leqslant t < 60$	8
$60 \leqslant t < 80$	7

a Copy and complete the cumulative frequency table below.

Time (t minutes)	Number of patients
$t < 10$	0
$t < 20$	3

b Draw a cumulative frequency diagram.
c Use your graph to estimate the number of patients treated for:
  i less than 30 minutes.
  ii less than 1 hour.
  iii between 30 minutes and 1 hour.
  iv more than 40 minutes.
d Use your graph to estimate the:
  i median.
  ii lower quartile.
  iii upper quartile.
  iv interquartile range.

6 Three silver and four orange hot air balloons (otherwise identical) were released at the same time.

a What is the probability that the first balloon to hit the ground is silver?
A prize was given to each child who could guess the colours of the first two balloons to hit the ground.
b Copy and complete the tree diagram of probabilities.

First balloon   Second balloon

c Use your diagram to calculate the probability that:
  i both balloons are silver.
  ii both balloons are the same colour.
  iii at least one of the balloons is silver.
d If this game were played 200 times, how many times would you expect a silver balloon to hit the ground first?

CHAPTER 13 Statistics 3

Practice

13A Statistical techniques

1. Criticise each of the following questions that were used in a questionnaire about pupils' music habits.

   a How do you listen to your favourite music?
         i MP3 player   ii CD player   iii Computer
   b How long do you spend listening in the evenings?
         i 0–1 hour   ii 1–2 hours   iii 2–3 hours   iv 3–4 hours
   c How many tracks of music do you have?
         i 0–500   ii 500–1000   iii 1000–1500   iv Other

2. A football team uses 21 players through the season. The number of matches each player has played in is given below.

   11, 9, 21, 19, 18, 18, 19, 22, 17, 18, 7, 3, 6, 11, 9, 13, 12, 18, 17, 11, 6

   a Use the data to copy and complete the stem-and-leaf diagram:

   ```
   0 |
   1 |
   2 |
   ```
   Key: | means ☐

   b Work out the median number of matches.
   c State the range.
   d How many players played more than the mode?

3. A pub quiz team is made up of people of different ages.
   The table shows the number of people in each age group.

Age group	Number
18–22	2
23–30	1
31–50	2
over 50	3

   a Represent this information in a pie chart.
   b Amy says: 'This shows that the same percentage of under 23s were chosen as 31 to 50s.' Explain why this statement might not be true.

4 Below are the scores of a class of 30 in a maths exam.

Boys	56, 17, 28, 87, 61, 90, 45, 37, 35, 48, 52, 58, 35, 9, 11, 78, 82
Girls	18, 93, 56, 75, 82, 55, 43, 49, 52, 28, 50, 44, 81

a Copy and complete the two-way table to show the frequencies.

	Boys	Girls
$0 \leq m \leq 20$		
$20 < m \leq 40$		
$40 < m \leq 60$		
$60 < m \leq 80$		
$80 < m \leq 100$		

b What percentage of the class scored over 60?
c What is the modal group for the girls?
d In which group is the median score for the boys?

5 The table below shows the weights of 100 dogs in a dog show.
a Copy and complete the cumulative frequency table.
b Draw the cumulative frequency graph.
c Use your graph to estimate the median and interquartile range.

Weight, w (kg)	Number of dogs	Weight, w (kg)	Cumulative freq
$0 < w \leq 5$	2	$w \leq 5$	
$5 < w \leq 10$	7	$w \leq 10$	
$10 < w \leq 15$	18	$w \leq 15$	
$15 < w \leq 20$	22	$w \leq 20$	
$20 < w \leq 25$	27	$w \leq 25$	
$25 < w \leq 30$	15	$w \leq 30$	
$30 < w \leq 35$	5	$w \leq 35$	
$35 < w \leq 40$	3	$w \leq 40$	
$40 < w \leq 45$	1	$w \leq 45$	

Practice

13B A handling data project

Investigate one of the following topics.

1 Compare the number of pages in books with the number of chapters.

2 Compare the number of digits that can be accurately remembered by different aged people.

FM **3** Investigate how accurately different people can estimate the weight of household objects. Compare different ages or male and female.

FM **4** Choose an investigation of your own.

CHAPTER 14 Geometry and Measures 4

Practice 14A Geometry and Measures revision

1 Find the area of each of the following shapes.

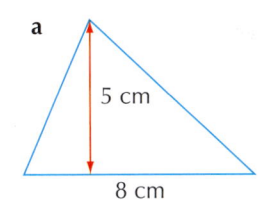

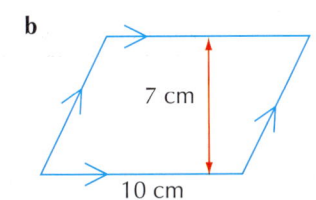

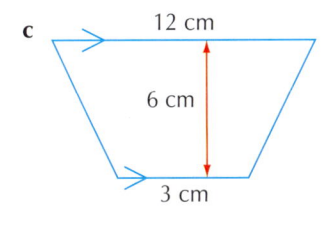

a triangle with height 5 cm and base 8 cm

b parallelogram with height 7 cm and base 10 cm

c trapezium with parallel sides 12 cm and 3 cm, height 6 cm

2 Calculate:
   i   the circumference and   ii   the area of each of the following circles.
Take π = 3.14 or use the π key on your calculator. Give your answers to one decimal place.

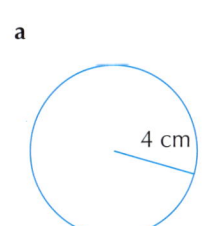

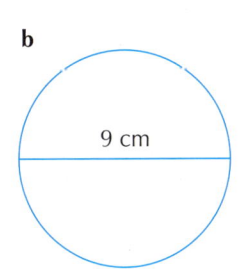

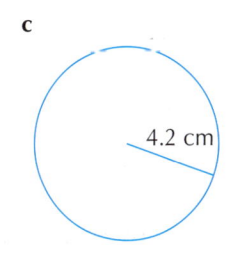

a circle with radius 4 cm

b circle with diameter 9 cm

c circle with radius 4.2 cm

3 Calculate: **i** the surface area and
ii the volume of each of the following 3-D shapes.

a
3 cm
10 cm
2 cm
4 cm
6 cm

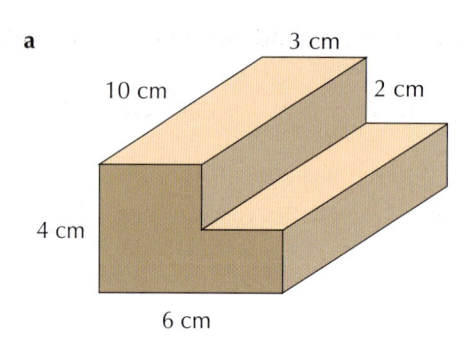

b
8 cm
10 cm
12 cm
6 cm

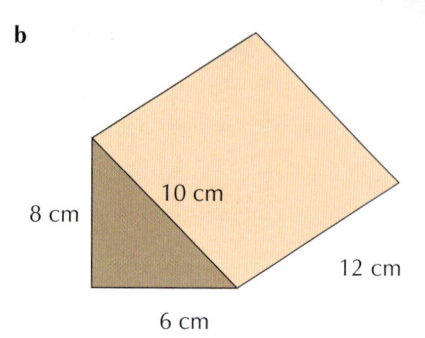

4 Calculate the volume of each of the following cylinders.
Take π = 3.14 or use the π key on your calculator. Give your answers to three significant figures.

a
3 cm
7 cm

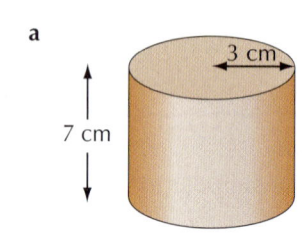

b
4 m
2.8 m

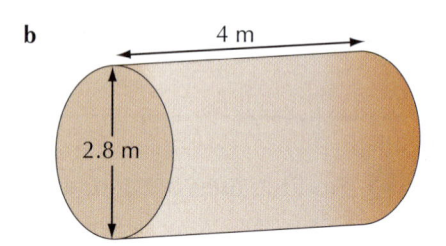

Practice

14B Geometry and Measures investigations

You are going to investigate tiling patterns.

Here are two square tiles, made from red squares and blue squares.

This 3 by 3 tile is made from 1 blue square and 8 red squares.

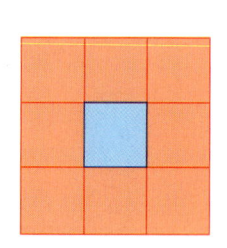

This 4 by 4 tile is made from 4 blue squares and 12 red squares.

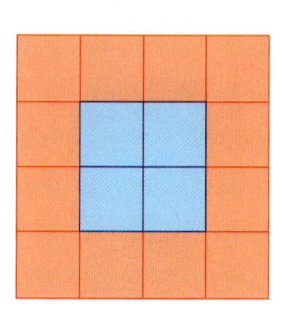

1. Draw diagrams to show how many blue squares and red squares are needed for 5 by 5 and 6 by 6 tiles?

2. Draw a table to show your results and write down any patterns you notice.

3. Can you predict how many blue squares and red squares are needed for a 7 by 7 tile?

4. How many blue squares and red squares are needed for an *n* by *n* tile?

Practice

14C Symmetry revision

1. Copy each of these shapes and draw its lines of symmetry. Write below each shape the number of lines of symmetry it has.

   a   b 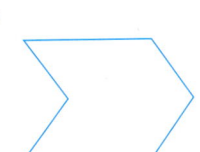   c 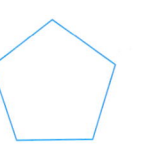   d 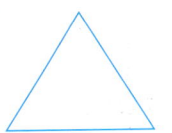   e

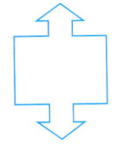

2. Copy each of these shapes and write the order of rotational symmetry below each one.

   a   b    c 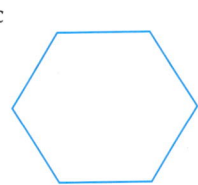   d    e

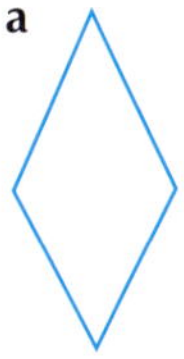

3. Write down the number of planes of symmetry for each of the following 3-D shapes.

   a Cube   b Cuboid   c Square-based pyramid   d Regular hexagonal prism

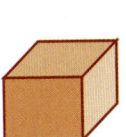

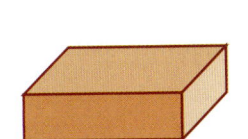

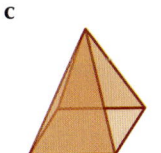

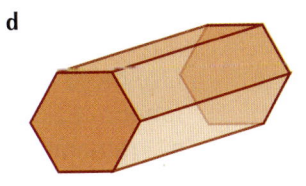

Practice
14D Symmetry investigations

You are going to investigate symmetry in regular shapes. Here is a regular hexagon with all its diagonals drawn in.

1 a How many different symmetrical shapes can you find inside the regular hexagon?

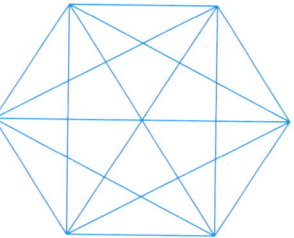

b On copies of the diagram, colour them in to make shapes which have either line symmetry or rotational symmetry.
  i Use two different colours at first.
  ii Then try using three different colours.

2 Now repeat the investigation by using a regular pentagon. Write down anything you notice.

CHAPTER 15 Statistics 4

Practice
15A Revision of probability

1 A scientist surveys fifty 14-year-olds, to see how many can remember a seven-digit number. He finds that eight of them can remember all seven digits, 15 of them can remember six of the digits.

Use these results to estimate the probability that a 14-year-old chosen at random has:
a remembered all seven digits.
b remembered six of the digits.
c cannot remember six or seven digits.

2 An eight-faced spinner is spun 100 times. Here are the results.

Number on spinner	1	2	3	4	5	6	7	8
Frequency	13	16	8	7	17	15	8	16

a What is the experimental probability of the spinner landing on the number 6?
b Write down the theoretical probability of a fair, eight-sided spinner landing on the number 6.
c Consider whether you think the spinner is fair.
d How many sixes would you expect from this spinner if it was spun 250 times?

3 The relative frequencies of the school buses arriving late one term are shown in the table below.

Number of days	5	10	15	20	25	30	35	40
Relative frequency of being late	0.6	0.5	0.6	0.55	0.56	0.6	0.6	0.575
Times late		3						

a Plot the relative frequencies on a graph.
b Explain why it is not possible to tell from the graph whether the bus was late on the first day of term.
c Write down the best estimate of the bus being late on any one day the following term.
d Copy and complete the table to show the number of times the bus was late.

4 Joy comes home one weekend with some maths and science homework. The probability that she get her maths homework all correct is 0.8, and her science all correct is 0.7.
What is the probability that in this weekend's homework:
a Joy has both maths and science homework all correct?
b Joy has neither of her homework all correct?
c Joy has at least one of her homework all correct?

Practice
15B A probability investigation

Carry out an experiment to investigate one of the following. Try to compare your experimental probability with the theoretical probability.

1 When you roll two dice, you get a total of 7 more times than any other total.

2 If you shuffle a pack of cards and deal out the top two cards, half the time these two cards will both be the same colour.

3 If you flip three coins, how often will you have at least two heads?

4 With a newspaper or magazine open in front of you, shut your eyes and, using a pencil, point it anywhere on the pages and see how often it actually lands on words, pictures or a space.

CHAPTER 16 GCSE Preparation

Practice 16A Solving quadratic equations

Solve these equations.

1
a $(x - 2)(x + 4) = 0$
b $(x + 7)(x - 6) = 0$
c $(x + 3)(x + 8) = 0$
d $(x - 5)(x - 5) = 0$
e $(x + 3)(x - 3) = 0$
f $(x + 4)^2 = 0$

2
a $x^2 + 12x + 35 = 0$
b $x^2 + 9x + 20 = 0$
c $x^2 + 9x + 18 = 0$
d $x^2 + 8x - 33 = 0$
e $x^2 - 14x + 13 = 0$
f $x^2 + 4x - 32 = 0$
g $x^2 - 14x + 48 = 0$
h $x^2 - 12x + 36 = 0$

Practice 16B Quadratic expressions of the form $ax^2 + bx + c$

1 Expand and simplify each of the following.
a $(2x + 5)(x + 3)$
b $(3x - 1)(2x + 3)$
c $(5x + 3)(x - 5)$
d $(4x - 3)(2x - 5)$
e $(3x + 2)^2$
f $(5x - 1)^2$

2 Factorise each of the following.
a $2x^2 + 5x + 2$
b $5x^2 + 27x + 10$
c $3x^2 + 17x + 10$
d $2x^2 - 15x + 7$
e $3x^2 + 2x - 5$
f $4x^2 - 4x - 3$
g $5x^2 + 18x - 8$
h $6x^2 - x - 2$
i $25x^2 - 10x + 1$
j $8x^2 + 15x - 2$

Practice 16C The quadratic formula

Solve these equations using the quadratic formula.

1 Solve these, giving your answers as whole numbers or fractions.
a $x^2 + 11x + 18 = 0$
b $x^2 - 8x + 15 = 0$
c $x^2 - 3x - 18 = 0$
d $2x^2 + 5x - 12 = 0$
e $x^2 + 8x = 0$
f $2x^2 - 18 = 0$

2 Solve these, correct to 2 decimal places.
a $x^2 + 5x + 3 = 0$
b $2x^2 + 8x + 5 = 0$
c $x^2 - 7x + 1 = 0$
d $5x^2 + 3x - 1 = 0$
e $4x^2 - 2x - 5 = 0$
f $2x^2 - 10x + 3 = 0$

3 Solve these, giving your answers in surd form, e.g. √3.
  a $x^2 + 5x + 1 = 0$
  b $2x^2 + 3x - 4 = 0$
  c $x^2 - 3x - 7 = 0$
  d $4x^2 + 3x - 3 = 0$

Practice
16D Completing the square

1 Complete the square.
  a $x^2 + 10x$
  b $x^2 - 24x$
  c $x^2 + 16x$
  d $x^2 - 10x$

2 Complete the square and simplify.
  a $x^2 + 4x - 4 = 0$
  b $x^2 - 6x + 3 = 0$
  c $x^2 + 2x - 2 = 0$
  d $x^2 - 10x + 23 = 0$
  e $x^2 - 4x - 6 = 0$
  f $x^2 + 14x + 19 = 0$

3 Solve these equations by completing the square. Give your answers in surd form where necessary.
  a $x^2 - 10x + 21 = 0$
  b $x^2 + 12x + 11 = 0$
  c $x^2 - 2x - 15 = 0$
  d $x^2 - 18x - 17 = 0$
  e $x^2 + 6x + 8 = 0$
  f $x^2 - 4x - 22 = 0$

Practice
16E The difference of two squares

1 Expand and simplify each of the following.
  a $(x - 20)(x + 20)$
  b $(2x - 7)(2x + 7)$
  c $(x - 5y)(x + 5y)$
  d $(3x - 2y)(3x + 2y)$

2 Factorise each of the following using the difference of two squares.
  a $x^2 - 169$
  b $x^2 - 10\,000$
  c $9y^2 - 49$
  d $x^2 - 16y^2$
  e $25y^2 - 1$
  f $36x^2 - 9y^2$
  g $100 - 81t^2$
  h $25k^2 - 4$

William Collins' dream of knowledge for all began with the publication of his first book in 1819. A self-educated mill worker, he not only enriched millions of lives, but also founded a flourishing publishing house. Today, staying true to this spirit, Collins books are packed with inspiration, innovation and practical expertise. They place you at the centre of a world of possibility and give you exactly what you need to explore it.

Collins. Freedom to teach.

Published by Collins
An imprint of HarperCollins*Publishers*
77–85 Fulham Palace Road
Hammersmith
London
W6 8JB

Browse the complete Collins catalogue at
www.collinseducation.com

© HarperCollins*Publishers* Limited 2008

10 9 8 7 6 5 4 3 2

ISBN 978-0-00-726807-8

Keith Gordon, Kevin Evans, Brian Speed and Trevor Senior assert their moral rights to be identified as the authors of this work.

All rights reserved. No part of this publication may be reproduced, stored in a retrieval system, or transmitted in any form or by any means, electronic, mechanical, photocopying, recording or otherwise, without the prior written permission of the Publisher or a licence permitting restricted copying in the United Kingdom issued by the Copyright Licensing Agency Ltd., 90 Tottenham Court Road, London W1T 4LP.

Any educational institution that has purchased one copy of this publication may make unlimited duplicate copies for use exclusively within that institution. Permission does not extend to reproduction, storage within a retrieval system, or transmittal in any form or by any means, electronic, mechanical, photocopying, recording or otherwise, of duplicate copies for loaning, renting or selling to any other institution without the permission of the Publisher.

British Library Cataloguing in Publication Data
A Catalogue record for this publication is available from the British Library.
Commissioned by Melanie Hoffman and Katie Sergeant
Project management by Priya Govindan
Covers management by Laura Deacon
Edited by Karen Westall
Proofread by Amanda Dickson and Tessa Akerman
Design and typesetting by Newgen Imaging
Design concept by Jordan Publishing Design
Covers by Oculus Design and Communications
Illustrations by Tony Wilkins and Newgen Imaging
Printed and bound by Martins the Printers, Berwick-upon-Tweed
Production by Simon Moore

Every effort has been made to trace copyright holders and to obtain their permission for the use of copyright material. The authors and publishers will gladly receive any information enabling them to rectify any error or omission in subsequent editions.

Mixed Sources
Product group from well-managed forests and other controlled sources
www.fsc.org Cert no. SW-COC-1806
© 1996 Forest Stewardship Council

FSC is a non-profit international organisation established to promote the responsible management of the world's forests. Products carrying the FSC label are independently certified to assure consumers that they come from forests that are managed to meet the social, economic and ecological needs of present and future generations.

Find out more about HarperCollins and the environment at
www.harpercollins.co.uk/green